The Natural Origins
of Economics

The Natural Origins of Economics

MARGARET SCHABAS

The University of Chicago Press CHICAGO & LONDON

MARGARET SCHABAS is professor of philosophy at
the University of British Columbia. She is the author of
*A World Ruled by Number: William Stanley Jevons and
the Rise of Mathematical Economics* and the coeditor of
Oeconomies in the Age of Newton.

The University of Chicago Press, Chicago 60637
The University of Chicago Press, Ltd., London
© 2005 by The University of Chicago
All rights reserved. Published 2005
Printed in the United States of America

14 13 12 11 10 09 08 07 06 05   1 2 3 4 5

ISBN: 0-226-73569-9 (cloth)

Library of Congress Cataloging-in-Publication Data

Schabas, Margaret, 1954–
    The natural origins of economics / Margaret Schabas.
    p. cm.
    Includes bibliographical references and index.
    ISBN 0-226-73569-9 (cloth : alk. paper)
    1. Economics. 2. Science. I. Title.
    HB74.S35S33 2005
    330.1—dc22

                                        2005014061

I dedicate this book to my son,
Joel Benjamin, with love and devotion.

CONTENTS

PREFACE

A question that has motivated much of my writing and research over the past twenty or so years is, In what sense is economics a science? This has necessitated a detailed study of the overlap—both methodological and conceptual—of economic theory with the natural sciences. My first monograph, which studied the rise of mathematical economics in late-nineteenth-century Britain, focused primarily on the methodological overlap, although the role of new concepts in logic and the philosophy of science was examined along with the appropriation of quantitative tools. The present study focuses on the conceptual foundations of classical economics and argues that the phenomena themselves were grounded in physical nature. It thus makes a very strong case for the natural context of economic theory from the early eighteenth century through the mid-nineteenth. John Stuart Mill stands out as the pivotal figure in the process that I refer to as the *denaturalization* of the economic order.

This book is much indebted to various funding agencies and foundations and to the input of numerous scholars. I wish to acknowledge the generous support of funds from the National Science Foundation, the Social Sciences and Humanities Research Council of Canada, the Faculty of Arts at York University, and the Faculty of Arts at the University of British Columbia. All four enabled me to reduce my teaching load and to uproot and undertake archival research at various libraries and institutions, including Duke University, Harvard University, the London School of Economics, Cambridge University, the National Library of Scotland, the University of Edinburgh, and the University of London. I also wish to acknowledge minor grants from the Wisconsin Alumni Research Fund, the National Research Council, and the Dibner Institute for the History of Science and Technology. These agencies helped me with travel and research.

It would be impossible to provide a complete list of the scholars who have given me verbal or written direction on this project, but I will do my best to acknowledge a subset. The largest subset is composed of my colleagues and students, both present and past. Thanks to them, I am reminded almost daily of why the life of the mind is to be cherished. There are some who merit special acknowledgment, as much for their wise counsel as for their detailed engagement with the specific arguments. Let me pay special tribute to Neil De Marchi, Mary Morgan, Carl Wennerlind, and Ted Porter. There are others who made it much easier for me to relocate and, thus, take advantage of libraries and collections outside Canada. Let me pay special tribute to Roy Weintraub, Daniel Kevles, Jed Buchwald, Jérôme Lallement, and James Secord. Then there are some who invited or encouraged me to publish in their volume or book and, thus, prompted me to hone in on some specific questions. Let me acknowledge (in addition to some already noted) Craufurd Goodwin, Arthur Fine, Donald Moggridge, John Davis, Philip Mirowski, Robert Nadeau, Bernie Lightman, Dorothy Ross, and Thomas Baldwin. Then there are those who have mentored me at some critical point in my career and, thus, helped me avoid various pitfalls. Let me acknowledge Scott Gordon, Trevor Levere, Sam Hollander, David Hollinger, Everett Mendelsohn, David Lindberg, Avi Cohen, Leslie Green, Margaret Morrison, Norton Wise, Annie Cot, and Lisbet Rausing. Finally, let me thank most wholeheartedly those who took the trouble to read chapters of the book: Sandra den Otter, John Beatty, Piers Hale, Ian Ross, and Leonidas Montes. Other participants in the process include André Lapidus, Philippe Fontaine, Loïc Charles, Anita Guerrini, Perry Mehrling, Malcolm Rutherford, Roger Emerson, Paul Wood, Michael Barfoot, David Raynor, Simon Schaffer, Geoffrey Harcourt, Tony Lawson, Andrew Skinner, Knud Haakonssen, Istvan Hont, Diane Paul, Paul Christensen, Harro Maas, José Luís Cardoso, Evelyn Forget, Jessica Riskin, Emma Spary, Mary Terrall, Vivienne Brown, Matt Strassler, Lisa Hill, and Jay Foster.

A special thanks goes to the late Susan Abrams, who recruited the manuscript and saw it to the stage of a signed contract prior to her early and tragic death from cancer. Her successor, Catherine Rice, has valiantly filled the breach and shepherded the book into print. Thanks to her and her associate Jennifer Howard for considerable patience and attentive assistance. Let me single out as well my research assistant, James Kelleher, for his persistent contributions toward improving the consistency and precision of this work.

Several segments of this book were published previously in journals or collected volumes. Parts of chapters 2 and 4 appeared as "David Hume on Experimental Natural Philosophy, Money, and Fluids," *History of Political Economy* 33 (2001): 411–35; © 2001 by Duke University Press, all rights reserved. Parts of chapter 5 appeared as "Adam Smith's Debts to Nature," in *History of Political Economy* 35, suppl. (2003): 262–81; © 2003 by Duke University Press, all rights reserved. Parts of chapter 7 appeared as "John Stuart Mill and Concepts of Nature," *Dialogue* 34 (1995): 447–65; and as "Victorian Economics and the Science of the Mind," in *Victorian Science in Context,* edited by Bernard Lightman (Chicago: University of Chicago Press, 1997), © 1997 by The University of Chicago, all rights reserved. Thanks to the editors and publishers for permission to reuse this material.

CHAPTER ONE

Before "the Economy"

The ideas of economists and political philosophers, both when they are right and when they are wrong, are more powerful than is commonly understood. Indeed the world is ruled by little else.

—John Maynard Keynes, *The General Theory of Employment, Interest and Money*

Daily references are made to *the economy,* whether in speech or in print. Most economists (and many politicians) maintain that virtually everything we do is governed by the economy. We are deemed to be producers or consumers at every moment of our lives. Every interaction can be defined as an act of exchange with an implicit price. Every object or service is potentially evaluated in terms of other goods and, thus, part of our so-called utility calculus. Even seemingly noneconomic activities such as marriage, suicide, or substance abuse have fallen under the lens of neoclassical economists, Gary Becker most famously. Paradoxically, just as we have come to see every human activity in economic terms, we have also come more and more to embrace the belief that we can control the economy, that it can be stabilized by monetary and fiscal measures. John Maynard Keynes's remark of economists and political philosophers—that "the world is ruled by little else" (1936/1973, 383)—has only grown in credibility and import.[1]

If one goes back a few centuries, however, it is by no means clear that people, even the learned communities of Western Europe, perceived such an entity as *the economy*. Collectivities such as markets and nation-states were well recognized, and "intricately interconnected" (Rothschild 2001, 29). But the majority who wrote on political economy defined their subject as an investigation into money, trade, and wealth; they rarely spoke of economies per se. *Commerce* connoted civil transactions between individual merchants and consumers and was not, by and large, used in the aggregate sense. Moreover, it was not just that early modern thinkers rarely referred to an economy, either national or international. Rather, their emphasis was much more on specific phenomena—the interest rate, for example—than on an overarching system by which the pieces fit together. Moreover, wealth was essentially a property of the physical world; the principles that governed its growth and distribution were said to be natural and could be augmented much as a forest might extend its reach into a meadow. Wealth was readily equated with the fruits of the earth and sea, with minerals, fish, and exotic plants. For Carl Linnaeus, the epitome of wealth, and, hence, the primary objective of the science of economics, was the domestication of foreign plants such as tea and cinnamon (see Koerner 1999, 2, 80, 150). At the end of time, humankind would restore the abundance and complete leisure of the Garden of Eden. Early modern conceptions of wealth were strongly wedded to prelapsarian ideals, whether one reads John Locke, Jean-Jacques Rousseau, or Thomas Jefferson, and were a far cry from more recent conceptions of wealth in terms of maximizing utility.

Only gradually, over the course of the late eighteenth century and the mid-nineteenth, did economic theorists come to posit and identify an economy as a distinct entity and maintain that it was subject, not to natural processes, but to the operation of human laws and agency. The main thesis argued here is that, until the mid-nineteenth century, economic theorists regarded the phenomena of their discourse as part of the same natural world studied by natural philosophers. Not only were economic phenomena understood mostly by drawing analogies to natural phenomena, but they were also viewed as contiguous with physical nature. Economic discourse was, in short, considered to be part of natural philosophy and not, as we would now deem it, a social or human science. It did not then address an autonomous sphere as it does today.

How and why political economists came to see the economic domain as severed from the physical world, as the product of human action, human deliberation, and human institutions, is the story imparted here. This *denaturalization* of the economic order, as I would like to call it, took place gradually and, in certain fundamental respects, has never been completed. But, if one were to juxtapose the conception of the economic realm articulated by leading theorists in the 1770s to those articulated in the 1870s, the differences are, I think, patently obvious. There is in both cases a strong conviction that economic phenomena such as money and trade are orderly and, thus, subject to the operation of laws, only the mechanisms by which those laws operate, and the source of the phenomena in general, stem from two different worldviews. For Adam Smith, nature was wise, just, and benevolent, whereas, for John Stuart Mill, it was imprudent, unjust, and cruel. For François Quesnay, wealth was a gift of nature, whereas, for Alfred Marshall, it could be defined only in terms of property claims and, thus, in terms of human institutions. For Quesnay, wealth was a physical entity, grain for our nourishment, whereas, for Marshall, it was a state of mental well-being or utility. For Smith, the best policy was to dismantle human designs and allow the "natural progress of opulence" (Smith 1776/1976, bk. 3). For Marshall, and even more for his student Keynes, the best policy was one of intervention, of fiscal management and economic planning.

Postwar economists, via manipulations of the interest rate and the money supply, have widely sought to stabilize the economy and dampen the oscillations of the business cycle. In short, there is now a sense that the economy can be engineered, if not entirely controlled. As a quick perusal of any macroeconomics textbook since the 1970s makes evident, economists are unequivocally committed to "stabilization policy." The only area for dispute is which "tools" to employ to achieve stabilization. Exogenous and often unpredictable shocks beset *the economy*, but, through fiscal and monetary measures, economic planners are able to steer us along a path of steady-state growth and stability. The rhetoric is very much like that of civil engineers. Wind shears or earthquakes or ordinary frost might threaten the stability of bridges, but engineers can overcome these shocks given the right plans. Macroeconomists convey a similar degree of confidence in our ability to achieve a national economy of low inflation and unemployment coupled with healthy economic growth.

For the eighteenth-century savants, there was no such separate domain. Facets of their world, manifest as trade, population, or wealth, received analysis, and specific measures such as taxation could be used to achieve specific ends, but there was no overarching concept of an economy as an integrated system of production, distribution, and consumption. There were goals, to be sure—perhaps to increase the national treasury as a means to wage war—but there was no collective thing to stabilize or engineer.

It is important to grasp just how recent is our configuration of an economy as an autonomous set of relations—only two hundred years old. Economic phenomena, by contrast, have been discerned and analyzed in their own right since antiquity. Aristotle's *Politics* and *Nicomachean Ethics,* for example, contain some remarkable analyses of money, market exchange, and household production.[2] Moreover, the terms *oikonomikē* or *oeconomia*, or the art of household management, and *chrēmatistikē,* or the more dubious activity of commerce, were established before Aristotle: arguably by Homer and most certainly with Xenophon's *Oeconomicus* of ca. 400 B.C. (see Booth 1993, pt. 1).

The term *oeconomy* became more prevalent in the early modern period, although more often in the context of natural history and physiology than in discussions of commerce and trade (see Schabas and De Marchi 2003). The *oeconomy of nature* became a commonplace term ca. 1700 and spawned many variants as well. Bryan Robinson, an Irish naturalist, published *A Treatise of the Animal Oeconomy* in 1734; François Quesnay issued his well-known *Essai phisique sur l'oeconomie animale* in 1736. Charles Bonnet appealed to an organic oeconomy to explain growth and generation in terms of the interaction of the different parts of an organism (see Daston 2004, 120). There were dozens of texts with similar titles throughout the mid-eighteenth century, even one, by the physician John Armstrong, entitled *The Oeconomy of Love* (1736). *Oeconomy* was also used to denote frugality or the wise management of one's household, but only in rare instances did it refer to the national economy at large. Jean François Melon's *Essai politique sur le commerce* (1734) uses the term but once, in the sense of a set of riches: "A better oeconomy produceth more Men, and a greater Plenty of all the Products of the Earth" (Melon 1734/1739, 54). Richard Cantillon had used the term once to refer to the arrangements of a slave plantation (see Cantillon 1755/1964, 33), and James Steuart's *Principles of Political Oeconomy* (1767), the title notwithstanding, makes only one reference to an oeconomy as we understand it (see Steuart 1767/1966, 1:16).

David Hume and Adam Smith rarely made reference to the *oeconomy* of a nation-state. In his *History of England,* Hume mostly used the term to mean individual frugality. In his *Dialogues concerning Natural Religion,* he used it in the sense of an orderly system, particularly in nature. Philo (Hume's spokesman) asks: "Is there a system, an order, an oeconomy of things, by which matter can preserve that perpetual agitation, which seems essential to it, and yet maintain a constancy in the forms, which it produces? There certainly is such an oeconomy" (Hume 1779/1947, 183). Philo extrapolates to "the whole oeconomy of the universe" (191) but also sees oeconomy in the structure of each animal (207) and even in the layout of a building (204). Only once, in his essay "Of Public Credit," did Hume remark on the "whole interior oeconomy of the state" (Hume 1777/1985, 354). Ironically, Smith used the term more in *The Theory of Moral Sentiments* than in *The Wealth of Nations.*[3] There are again references to the *oeconomy* of a person, meaning his frugality, but more often the word appears as the *oeconomy* of nature (see Smith 1790/1976, 50, 77–78, 183, 321).

These are all signs that, until the late Enlightenment, the natural and the economic realms were one and the same. Even in the 1790s, Goethe claimed that "nature was the perfect economy" (see Jackson 1994, 411). Concepts and methods from natural history and the physical sciences shaped and governed the analyses of wealth offered by eighteenth-century thinkers such as Hume, Quesnay, and Smith. And none had a concept of an economy as we denote it at present or even as Ricardo devised, namely, a self-contained and self-regulating system of the production and distribution of commodities that can be more or less efficient and that can grow or decline. Indeed, as we will see in the next chapter, the most detailed account of an oeconomy during the mid-eighteenth century was the Linnaean oeconomy of nature, which encompassed human production, the web that joins animals and plants, the earth's surface (including fossils), and what we would now call the *hydrologic cycle* (see Worster 1977, 34–35).

For Smith, moreover, political economy as a pursuit was defined as the "science of the legislator," not a full-fledged study of economic phenomena (see Winch 1996, 409). Only in the first half of the nineteenth century did the concept of the economy emerge as an autonomous entity. John Stuart Mill was pivotal in rendering explicit the role of human agency as the framework for economic analysis, as the preliminary remarks to his *Principles of Political Economy* reveal: "In so far as the economical condition of nations turns upon the state of physical knowledge, it is a subject for the physical sciences, and the arts founded on them. But in so far as the causes are moral or psychological, dependent on institutions and social relations, or on the principles of human nature, their investigation belongs not to physical, but to moral and social science, and is the object of what is called Political Economy" (Mill 1871/1965, 1:20–21). Mill thus captures the very transition that took political economy from the physical to the social domain.

Several caveats need to be issued. For one thing, the term *social science* was coined only in the 1780s, by Condorcet, and did not enter the English language until the early nineteenth century (see Baker 1964; and Cohen 1994, xxvii). The factors that led to the emergent field of social science were part and parcel of the process by which economics detached itself from natural philosophy. For another, economic theorists have always embraced specific notions, either explicit or implicit, about human nature. I do not wish to imply for a minute that such notions were absent in the early modern period. Rather, they were refracted through the prism of physical nature, at least in the eighteenth century and the early nineteenth. While it would be imprudent to define *nature*— clearly the term has multiple meanings even at a given point in time—I will most often use the term to mean physical nature, the domain commonly studied by natural philosophers at the time under consideration. The phenomena under investigation might belong to the science of mechanics, or to experimental physics, or to natural history. For the sake of my argument, human nature, or the phenomena known as *the will, the intellect, the sentiments,* and *the passions,* is to be set apart from the rest of the natural world, even though there are some obvious points of overlap, human physiology most notably. I propose that, while human agency was not emphasized as the primary or proximate cause of economic phenomena in the mid-eighteenth century, that notion is, nonetheless, present in economic writings of the period.

Hume wisely remarked that no word "is more ambiguous and equivocal" than *nature* (Hume 1739–40/2000, 304). The passage of 250 years has done nothing to improve the situation. Raymond Williams commences his efforts at definition with the caveat: "Nature is perhaps the most complex word in the language" (Williams 1976, 184). It is not just the ambiguity and complexity of the word *nature* that make it one of the most elastic and evasive terms in our language but the wide array of definitions. Familiar distinctions—such as those between *natural philosophy* and *moral philosophy, physical nature* and *human nature, nature* and *artifice, the natural* and *the supernatural*—are, ultimately, drawn in sand, if only because no two philosophers have ever ascribed exactly the same meanings to those terms.[4] Williams rightly distinguishes one enduring set of meanings that maps nature onto the material world, but it is an open question where to place human beings. Even if one were to separate human agency from physical nature, in a manner embraced by modern science, there is, perhaps, no place, at least in the sublunar realm, that is untainted by human activity. The ozone layer, Bruno Latour has argued, is a political object; the distinction between nature and society, he argues, is completely devoid of meaning (Latour 1993). It is not just that everything that exists is part of nature but that everything is also part of the social realm. As Lisbet Rausing has recently observed: "In its full brittle complexity, nature exists only on our sufferance" (Rausing 2003, 173).

Although it is almost impossible to articulate precisely what is meant by the word *nature* in the period under consideration here, it is possible, looking to paintings or poetry, for example, to discern from tableaux or portraits of nature the range of credible meanings. Broadly speaking, for the Enlightenment philosophers, physical nature was rational, harmonious, and orderly. They frequently spoke metaphorically of nature as a watch, a clock, or, more generally, a machine. Some—for example, Bernard de Fontenelle and Adam Smith—viewed nature as a kind of theater, by which elaborate hidden machines produce delightful results (see Schaffer 1993, 493). As Lorraine Daston has observed of the eighteenth-century metaphysicians: "The natural no longer subsumed the artificial, as it had for Bacon and Descartes; rather, the artificial subsumed the natural" (Daston 1998, 166). For the Romantic philosophers of the early nineteenth century, nature had acquired a mysterious and turbulent appearance.[5] Nature was shrouded in layers of manifest symbols that, rightly discerned, would reveal its underlying unity. For many mid-nineteenth-century philosophers, nature had become fierce and almost diabolical, something to be conquered and subdued.[6] Think of Darwin's depiction of nature as a continual battlefield between weak and strong, "red in tooth and claw," to use Tennyson's apt phrase (see Ruse 1979).

Another avenue to conceptions of nature in this period is that of gardens, particularly aristocratic gardens. As Chandra Mukerji has argued, French formal gardens of the ancien régime were symbols of corporate property, of mercantile capitalism, rather than of paradise or unbridled fertility (Mukerji 1993, 456). Alternatively, the English landscapers "provided a model of natural perfectibility" to incite improvement (453). Paintings and gardens, among other objects, offer a broad spectrum of images of nature devised during the eighteenth and nineteenth centuries. Even metaphysicians such as Fontenelle or George Berkeley embedded their discourse in dialogues conducted in gardens, as a means of reminding the reader of the visual cues to the order of nature. To this day, *nature,* as Williams observes, has also meant the countryside or places unspoiled by humans (Williams 1976, 188).

The eighteenth century also abounds with treatises on the nature of nature. Two salient examples are Dénis Diderot's *De l'interprétation de la nature* (1753) and Baron d'Holbach's *Système de la nature* (1770), but there are dozens more such tracts. As Carl Becker observed, the words *nature* and *natural law* were to the eighteenth century what *evolution* and *progress* were to the nineteenth.[7] A century later, philosophers were more cautious about making lengthy pronouncements or did so under the guise of an overarching system such as Herbert Spencer's. By the mid-twentieth century and the onset of academic specializations, most accounts had become purely historical, making sense of past ideas: pace the sweeping and splendid accounts of Arthur Lovejoy (1936) and Robin George Collingwood (1945). Among postwar scholars, modesty has rightly set in, and one is hard-pressed to find more than a few general accounts on the nature of nature. The 1989 Herbert Spencer lectures at Oxford elicited six general essays on the concept of nature by leading scholars such as Geoffrey Lloyd, Elliott Sober, and Roger Penrose, but each essay sticks fairly closely to the author's respective subdiscipline (see Torrance 1992). Kate

Soper's *What Is Nature?* (1995), while more substantial, is leveled at environmentalists and their misguided appeals to nature. Lorraine Daston's two recent articles, "The Nature of Nature in Early Modern Europe" (1998) and "Attention and the Values of Nature in the Enlightenment" (2004), though replete with apt comparisons and contrasts between the early modern period and our own, are judiciously positioned in the past rather than the present.

Collingwood's *Idea of Nature* maintained: "Greek natural science was based on the principle that the world of nature is saturated or permeated by mind. . . . The world of nature is not only alive but intelligent; not only a vast animal with a 'soul' or life of its own, but a rational animal with a 'mind' of its own" (Collingwood 1945, 3). This was especially true for the Stoics. To achieve greater virtue was part and parcel of being in harmony with physical nature, to understand that the *pneuma* or spirit pervades all bodies. While Collingwood is overblown and dated, there is still a large grain of truth to his central thesis. As numerous scholars have since argued, the ancient theories of natural philosophy, atomism, essentialism, and the like were directly linked to theories of virtue and justice. In that respect, the sphere of human activity, such as the *polis,* was enmeshed with that of physical nature (see Lloyd 1992, 15).

Scholars have also placed much emphasis on the strong intellectual debts of the leading Enlightenment economists—Hume, Smith, and Quesnay—to Greek philosophy. Both Hume and Smith were enthusiasts of Stoic thought, while Quesnay attempted to revive Aristotelian and Galenic ideas. In this respect, although also under the sway of the mechanical philosophy that swept over European thought, they were still deeply wedded to the supposition of a unified world, the issue of a single author or deity. The separation of mind and matter, in short, did not fully seep into political economy until the nineteenth century.[8]

These are grandiose claims, claims that surely demand extensive support from the historical record. In addition to a close reading of the primary sources in print, I have perused some of the manuscript collections at Harvard University, the University of Toronto, Cambridge University, the University of Edinburgh, and the National Library of Scotland. My efforts have been facilitated by considerable scholarly spadework already extant; my debts to those who have edited and compiled the papers and correspondence of the leading past thinkers are incalculable. I also rely extensively on the secondary literature in the history of science, which has gained considerable sophistication in the postwar period and grown at a dramatic pace.

Twenty-five years ago it was a truism among historians that eighteenth-century science was enigmatic. There were brilliant studies of Enlightenment physics by Charles Gillispie (1960), Robert E. Schofield (1970), and Thomas Hankins (1970). But they were much overshadowed by scholarship of seventeenth- and mid-nineteenth-century science, partly because scholars were drawn to such luminaries as Galileo and Newton, Darwin and Maxwell. The eighteenth century does not have a single contributor to science with such cachet. Indeed, the period rightly belongs to the moral sciences, or *the science of man*, as it was known. The relative political stability of the eighteenth century oddly enough engendered brilliance in political philosophy—Rousseau, Hume, and Smith, to name just a few. And, while it was possible to study some of the prominent mathematicians—such as Leonhard Euler—in isolation from social thought, it was not so feasible in the case of D'Alembert, Lavoisier, Priestley, or Condorcet. Arguably more than historians of seventeenth- or nineteenth-century science, those of eighteenth-century science are compelled to be versed in political philosophy. Some of the pathbreaking studies that take an integrated approach to Enlightenment science are those by Baker (1975), Gillispie (1980), and Daston (1988). More recently, the number of detailed studies of the culture of science in the eighteenth century, both narrowly focused and more wide-ranging, has grown considerably (see, e.g., Golinski 1992; Koerner 1999; Riskin 1998; Schiebinger 1993; Schaffer 1983, 1989, 1990, 1993, 1999; and Spary 2000). It is now possible to make some reliable generalizations and, thus, begin to position economic thought within this context. There is far less literature, however, that treats economic ideas within the context of the history of science. This book is meant to provide a contribution to that end, and the result, I trust, will be to impress the reader with the great number and wide variety of links between classical economics and the natural sciences, both conceptual and methodological.

Historians of economics have traditionally adopted an internalist stance and emphasized what Emma Rothschild aptly calls *processionalism,* by which she means viewing the past as a set of unadulterated seeds for present truths (Rothschild 2001, 40–41). Such tendencies stem partly from the strongly ahistorical nature of advanced training in economics; the science is studied and practiced in a way that rarely cultivates a historical sensibility.[9] But even more significant is the tenor of Joseph Schumpeter's monumental *History of Economic Analysis* (1954), which looked to every past thinker as either aiding or delaying the onset of neoclassical economics. Schumpeter, however, was not strongly internalist. His book makes a concerted effort to document the broader historical context notwithstanding the primary emphasis on analytic progress. Fortunately, in the past fifteen or twenty years, a number of prominent historians of economics have broken with this tradition and challenged the received view of a unilateral and purportedly triumphant march toward greater analytic clarity. Even more exciting has been the significant number of works that join the history of economics to the history of the natural sciences, by specialists in both fields.[10]

My intentions are not to rework the internal economic analyses that one finds in the leading texts of Smith or Mill, let alone translate past theories into contemporary mathematical language, as has been done by many other scholars. Rather, I wish to appraise the claims of the leading classical economists in relation to the scientific ideas and practices of their time. I seek zeitgeists, not systems of simultaneous equations.

Classical economics is about as disputed a historical term as any other. Some would stretch its purview from Petty and Locke right up to Marx, while others would limit it to the one hundred years that run from Smith to John Elliott Cairnes.[11] Not much is at stake with either set of boundaries. I use the term *classical economics* simply as a shorthand for a body of ideas that spans the period from the mid-eighteenth century to the mid-nineteenth. I begin my study of economic ideas with the French *économistes* of the eighteenth century because they, more than any other group, so readily joined their conception of wealth with the operations of nature. The story might have begun at an earlier point, but there are no stronger instantiations of economic thought harnessed to nature than the Physiocrats.[12] Chronologically, Hume's economic essays of 1752 preceded the formation of Physiocracy by five years, and the heyday for the latter was truly in the 1760s. Moreover, if one grants that some ideas central to Hume's political economy are in his *Treatise of Human Nature* (1739–40), then Hume ought rightly to be treated before the Physiocrats. The reason for my reverse ordering has to do with Hume's close ties, personal and intellectual, with Adam Smith and the fact that Hume segues well into Smith. Moreover, the Physiocrats are commonly placed apart from the school of classical political economy (Hume's status is also ambiguous), and, thus, for that reason too, it seemed best to commence with a study of their ideas.

To develop my reading of this period as one of denaturalization, I might well have started even earlier. Hobbes, Boisguilbert, Petty, and Locke all stand out as relevant figures in this account. But an additional reason for starting in the mid-eighteenth century, at least in a concentrated fashion, stems from my belief that political economy was not viewed as a separate and coherent discipline until that point in time. The seventeenth century abounds with books, pamphlets, and broadsides on money, interest, and trade, to take the title of Locke's 1696 volume as representative. A discourse can easily be located ca. 1690, but the subject of political economy did not form a domain in its own right until the mid-1750s. The mid-eighteenth century witnessed the founding of chairs and societies in political economy (notably in Sweden and Italy), the establishment of periodicals on economics (especially in France), and the publication of general treatises such as Richard Cantillon's *Essai sur la nature du commerce en général* (1755). Hume's economic essays of 1752 circulated widely. Within two years, there were two new English editions and two French translations. By 1770, these essays had been reissued at least seventeen times and translated into five European languages (see Charles, in press).

As for the end point of the period of classical economics, I take Mill and Cairnes rather than Marx. As Dennis O'Brien points out, Marx's economics are borrowed wholesale from the classical economists, but he is an offshoot, like Henry George (see O'Brien 1975, xi). To make sense of Marxian economics requires attention to Hegel and the French socialists as much as to Ricardo and Mill. Moreover, Marx's major work, *Das Capital* (vol. 1 appeared in 1867) did not begin to make its mark until the 1880s, when my account tapers off. Nevertheless, there are many linkages between nature and economic phenomena in Marx's works, not to mention an essentialism regarding our species-being that drives much of his analysis (see, e.g., Hearn 1991; and Meikle 1991). And, while I am partial to the view that Mill and Cairnes were the last of the classical economists, I will also underscore here the extent to which Mill stands as an important pivotal figure for the neoclassical era.[13] Indeed, if there is any writer who serves to mark the denaturalization of the economic order, it is Mill—and the Mill of the 1836 *Essays* as much as of the 1848 *Principles*. In that respect, my overview is confined to about a hundred years, from approximately 1750 to 1850.

What makes the Enlightenment conception of nature so different from our own is its strong allegiance to the deity. The moral and natural worlds were unified insofar as they were the book of a single "Author," as philosophers then were wont to say. Only a small minority of philosophers, such as Hume, queried the view that a deity had made the world according to a given plan. More significant was the belief in the continuous presence and providential action of the Christian God. Despite secularizing forces in the late Middle Ages, many seventeenth-century savants intensified and extended the identity of God with physical nature. This is best exemplified by Nicolas Malebranche and his doctrine of occasionalism, whereby God was the proximate cause of all events. Insofar as all matter was passive and without soul, it could not be active. God was the only source of motion and change, and, hence, every mechanical event bespoke his agency. Over the next century, God's efficacy begin to wane, although never to the point of robbing nature of its order. The secularization of concepts of nature transpired at a slow pace and was by no means a simple or monotonic development. Voltaire, Hume, Buffon, and Hutton were some of the prominent figures at the vanguard of secularization. The nineteenth century brought even more and profounder uncouplings of science and religion, thanks in part to the maturation of evolutionary biology in the hands of Cuvier, Lamarck, Lyell, and Darwin. But there were also many others—Whewell and Pasteur come to mind—who were bent on reinforcing the Christian component of their scientific research. The secularization of science will be examined in more detail in the next chapter. The main point to grasp here is the extent to which Enlightenment philosophers, like their Greek predecessors, saw the world as a single unified whole.

Contemporary natural and social scientists, by contrast, have implicitly agreed to divide the world into two parts. When physicists today think of the world they investigate, it is one with all the social institutions stripped away. Matter, force, and energy are studied independently of the principles that govern interest rates or property laws, let alone the accumulation of wealth. Economists, in parallel fashion, have come to adopt a domain of discourse that is similarly segregated. The Schroedinger equation can tell us nothing about the interest rate, to put it bluntly. To be sure, some of the analytic tools resemble one another; there are many formats common to the models used in microeconomics and ecology, for example.[14] But there is no ontological conviction that economic phenomena are themselves part of the natural order that serves as the domain for natural scientists. The phrase *laissez-faire* has lost all its original import. It pertains now to liberal trade policies, not the operations of nature itself.

Economists today study a world that is essentially detached from the processes of physical nature. The annual harvest that figured so prominently in the classical theory of political economy has all but disappeared. True, some calendar events, such as the Christmas retail market, have economic consequences, but those consequences stem from human convention rather than from "natural processes." The cycle of the seasons that was at the forefront of economic theory ca. 1800 has now almost disappeared from the discourse. Indeed, economists tend to mock the idea once put forward by Stanley Jevons that the sunspot cycle might influence the business cycle via the climate. If anything, fluctuations in the climate are rarely present in economic models and can, arguably, be eliminated from the model, given the right technology.[15]

My thesis of the denaturalization of the economic order is not intended to undercut the extent to which economists have persistently sought to emulate physicists and, thereby, embraced mechanical analogies and mathematical tools. Beyond doubt, contemporary economics bears a strong formal resemblance to physics, as Philip Mirowski has ably demonstrated in his *More Heat Than Light*. There is remarkably little effort made, however, to link the uniformities of the physicist's world to those of the economist's. Indeed, even the physical description of economic processes such as production is very meager. As Mirowski has noted: "The symmetry of consumption and production has severely limited and crippled any pretense of cogent description and discussion of actual physical production processes. . . . The problem of production has been reduced to a matter of semantics" (Mirowski 1989, 319). Despite all the analogical and methodological similarities in their respective discourses, contemporary economists and physicists inhabit separate worlds.

Jevons was the first to explore the question of the dimensions of economic variables, and he found that, in many cases, the physical component dropped out of the analysis. Demand and supply, for example, were measured in terms of utility, which in turn was measured in terms of intensity and time. The upshot of this is that more contemporary neoclassical notions, such as commodity space, are entirely ethereal. They have no grounding in physical nature as it is commonly understood. Indeed, the material attributes of goods are of no consequence to the analysis; all that matters is that commodities map onto utility functions under the requisite constraints. In fact, were such goods to become corporeal, it would rob the indifference curves and, hence, the demand functions of their continuity and completeness and, thus, make the use of the differential calculus all the more unwarranted (see Hausman 1992, chap. 1).

Other attributes have also lost their physical dimensions. Money, for example, is now seen as the locus of information about present and future prices (see Brunner and Meltzer 1971). Capital is essentially a claim on the future and is, thus, defined in terms of the discounting of time. Perhaps the only variable in neoclassical economics that might still be ascribed a physical dimension is population growth. But, even there, economists have stripped it of its natural dimensions. For Malthus, offspring were the result of a natural and enduring passion between the sexes. For Gary Becker, they are the result, not of a reproductive urge, let alone of sexual passion, but of an explicit component of our deliberations or utility calculus. Now all children are planned like the purchase of consumer durables such as refrigerators; they offer a stream of benefits after an initial hefty investment (see Becker 1976, 171–94). Even more unorthodox is the assumption, widely used for mathematical closure, that we live infinite lives—insofar as our accumulated wealth is bequeathed to our offspring. Thomas Mann had it right in his masterful *Buddenbrooks* when he observed that children are the only claim that rich burghers have to immortality.

With the advent of neoclassical economics, we find a pronounced shift toward a conception of the economy that stems from the deliberation of economic agents. As Jevons declared in 1871: "The theory presumes to investigate the condition of a mind, and bases upon this investigation the whole of Economics" (Jevons 1871/1957, 14–15). Insofar as prices were construed as the direct product of a Benthamite calculus of pleasure and pain, and insofar as prices were conceived as the simultaneous clearing of production and distribution, the entire economy emanated from individual deliberation. As Philip Henry Wicksteed observed in an entry for *Palgrave's Dictionary of Political Economy* (1896): "The economist must from first to last realise that he is dealing with psychological phenomena, and must be guided throughout by psychological considerations" (Wicksteed 1910/1933, 767). There was, to be sure, much human agency in the classical theory, but it did not stem from what we would now call *rational choice*. Human nature was infused with certain passions and propensities—to better our condition or to truck, barter, and trade, for example—but the rest of the social fabric unfolded with unwitting regularity against a backdrop of an orderly world. It is not by accident that Smith wrote on the history of astronomy or compared the activities of the market to gravitational attraction.

As Maurice Dobb recognized several decades ago, the classical theory also took classes, not individuals, as its unit of analysis (Dobb 1973, chap. 7). There were different types of persons (frugal merchants, prodigal sons, etc.), not to mention the standard classes of farmers, landowners, and artisans, but, within these groups, there was little to distinguish one person from another. Smith's *Wealth of Nations* is full of remarks that confirm this preoccupation with groups:

> Masters are always and everywhere in a sort of tacit, but constant and uniform combination, not to raise the wages of labour above their actual rate.

> People of the same trade seldom meet together, even for merriment and diversion, but the conversation ends in a conspiracy against the publick, or in some contrivance to raise prices. (Smith 1776/1976, 1:84, 145; see also Denis 1999)

There was such regularity in human action, at least among types of persons, that the role of individual deliberation paled in comparison to the role of the group, at least for eighteenth-century economics. As David Carrithers has observed for thinkers such as Adam Ferguson: "Individuals qua individuals . . . are not the building blocks of society. Rather, families, clans, and socioeconomic groupings of various sorts deriving from the earliest division of labor began to take center stage as the focal point of the 'sociological' analysis emerging in the Enlightenment" (Carrithers 1995, 235).[16] Similarly, Bernard Mandeville served to show that "the individual's point of view, while instinctive in naturally self-regarding creatures, in fact tends often to conceal the social significance of his actions" (Hundert 1994, 179). People are deceived into thinking that they act only as individuals when, in fact, they are entirely the products of their social milieu.

So deeply entrenched is our commitment to individuals as the unit of analysis in economics that it is hard to accept that such a commitment played a much lesser role in the early modern period. There were, certainly, efforts to probe the mind of a single person, as Smith does with his image of the impartial spectator. But, again, that spectator makes us more than one; it joins us to a group, a collective mind of moral beings.[17] Hume has already shown that the individual's boundaries dissolved under scrutiny, that there was no receptacle holding together the stream of mental impressions that seemed to constitute and individuate the mind of a single person (see Baier 1991). Other prominent Enlightenment philosophers—think of Rousseau's general will or Kant's categorical imperative—shared this propensity toward the collectivity rather than the individual.

In neoclassical economics, by contrast, economic laws emerge precisely because of the distinct calculations of different minds, different preference sets for goods and services, and different attitudes toward risk and toward time. Given the number of variables and the range of preferences, no two persons are the same, and, hence, groups of individuals rarely figure in theoretical analysis. It is the very diversity of persons, particularly our mental stock of beliefs, desires, and intentions, that accounts for the existence and pattern of economic activities.[18] Indeed, we are so unique that no two of us use the same means to define or measure our utility. The inscrutability of each mind precludes interpersonal comparisons of utility.

Our current conception of the economy is not just closely wedded to methodological individualism but mind driven through and through (see, e.g., Davis 2003). We all seek to maximize our utility, however that may be construed or measured. This in turn raises significant and possibly intractable problems at the explanatory level, particularly insofar as the source of such diversity is located in the realm of the unobservable. The traditional move since Paul Samuelson has been to slay that metaphysical beast by retreating to the stance of revealed preferences. Our consumption bundles reflect or "reveal" our preferences, period. But, as several critics have shown, neoclassical theory incorporates an account of mental deliberation even if it chooses not to address it in a substantive manner (see Hausman 1992, 19–22). Economics is still completely beholden to the operations of the mind.

Fortunately, the early neoclassical economists were not so bashful about the concept of utility. For Francis Ysidro Edgeworth, economics was virtually synonymous with the utility calculus (Edgeworth 1881/1967, 15–16). For Alfred Marshall, all that man can produce or alter are utilities: "Man cannot create material things. In the mental and moral world indeed he may produce new ideas; but when he is said to produce material things, he really only produces utilities; or in other words, his efforts and sacrifices result in changing the form or arrangement of matter to adapt it better for the satisfaction of wants" (Marshall 1890/1920, 53). Matter no longer constrains or determines, as it did for the Physiocrats and the classical economists. No longer does land scarcity or the stationary state loom on the horizon. We find a conception of the economic order that is more or less severed from physical constraints. Wealth, or utility, is granted an unprecedented ability to expand.

What brought about this transformation that I am calling the *denaturalization* of the economic order? Clearly, many factors were at work. In addition to the internal reworking of economic concepts, I will also explore related developments in natural philosophy itself, experimental physics, natural history, and evolutionary biology as well as in the emerging science of the mind. The ongoing secularization of science also played a significant role in the shifting convictions of eighteenth- and nineteenth-century economists. This will be a more difficult set of links to establish—but relevant nonetheless.

Developments in the economy itself no doubt made their mark as well. Industrialization is the salient feature of the period under consideration here, particularly in Britain and France. Maxine Berg has argued that the "age of machinery" was reflected in the economic texts of Ricardo and his immediate successors (Berg 1980). Likewise, with increasing urbanization, it might seem that appeals to nature, at least in the form of agriculture, would fade into the background. This has a very plausible ring to it, but there are problems with such claims. For one, economists often help themselves to phenomena from across the globe and as far back as historical records permit. It might seem obvious that, prior to the onset of industrialization, the Physiocrats would put so much stock in the agrarian sector or that, given the Great Depression, John Maynard Keynes would highlight the problem of unemployment, but these phenomena were neither new nor outside economists' sphere of observation at prior historical moments. To explain why one economist or a group in a given time and place embraces a select set of ideas requires much more than an appeal to extant economic conditions. In my view, even for someone as worldly as Keynes, his theoretical insights were less a product of the economic slump than one might suppose.[19] An economist is much more inclined to work out his or her ideas in the confines of the study, absorbing the books and papers of others. The more we know about Keynes's conceptual genesis, for example, the more we see that it was the product of philosophical ideas and the effort to sort out inconsistencies in the work of his immediate predecessors (see Davis 1994; Bateman 1996; and Runde and Mizuhara 2003). Arguably, a better case could be made for the claim that the successful reception of economic theories is more likely when the theoretical claims coincide with similar economic conditions. In sum, historical inquiries often disclose a fair degree of autonomy between economic theorizing and the economic conditions experienced by the theorist.

My central claim—that the concept of an economy is effectively a post-Enlightenment one—may be found in Michel Foucault's *The Order of Things*. He there argued that, ca. 1800, three branches of knowledge—political economy, philology, and biology—all emerged in the gaps created by three previous epistemes, the studies of wealth, grammar, and natural history (Foucault 1970, 207). His thesis thus implicitly commits him to deeper forces at work in intellectual history, such that three different "human sciences" emerged simultaneously because "representations" no longer offered satisfactory explanations. However appealing, the account remains descriptive, without much guidance as to why these sciences moved in step with one another.

In my work here, I am not directly concerned with whether political economy evolved in tandem with other human sciences. It seems to me highly unlikely that it did, if only because the discipline of economics has a much older lineage than sociology, linguistics, or even psychology. As Joel Kaye (1998) has recently shown, there is a rich body of literature on money and trade in the fourteenth and fifteenth centuries that closely parallels investigations in natural science. Aquinas, Jean Buridan, and Nicole Oresme all took stock of new developments in European commerce in conjunction with Aristotle's teachings. But, to return to Foucault's specific treatment of economics, it seems too that he misleads with his claim that political economy came into being only with Ricardo. In my view, it was already a coherent discourse by the late seventeenth century, particularly given the texts of Pierre Boisguilbert, John Locke, and William Petty.[20] To be fair, Foucault displays considerable erudition about the history of economics. He acknowledges the importance of Petty, Boisguilbert, and Ferdinando Galiani, for example. But I do not agree with his claim that the concept of wealth was a representation in the seventeenth and eighteenth centuries. Rather, it was conceived of in hard physical terms. Only later, in the mid- to late nineteenth century, did it become a nonmaterial entity and, thus, if I understand Foucault correctly, a representation. So, while I agree with him that Ricardo provides a critical watershed, I stand Foucault on his head. Wealth went, not from being a representation to an object, but the other way round.

Foucault also claims that "wealth is a system of signs that are created, multiplied, and modified by men; the theory of wealth is linked throughout to politics" (Foucault 1970, 205). This too is misleading. For many seventeenth- and eighteenth-century writers, say Locke, Quesnay, or even Smith, wealth was created prior to anything political, or so they believed. The strongest case is Locke's state of nature, which has property and money prior to the formation of government. Wealth was an inextricable part of physical nature and, thus, subject to physical laws. Human institutions, including the political system, were to be considered separately and post facto, at least until the mid-nineteenth century.[21]

A leading scholar who commits to a pre-nineteenth-century concept of the economy is Catherine Larrère. Her *L'invention de l'économie au XVIIIe siècle* (1992) contains a very compelling interpretation of the Physiocrats and of Turgot. She emphasizes the deep rift with the mercantilist thinking of Vincent de Gournay and Josiah Child and the strong appeals to physical nature and natural law that motivated Quesnay and his followers, including their aspirations to scientific thinking. But I take issue with her claim that the Physiocrats invented the concept of an economy. For one thing, there is no such referent in the Physiocratic literature, and, for another, the Physiocrats' very efforts to understand the circulation and reproduction of wealth are explicitly articulated in terms of a natural order. In short, there is no autonomous economic order to be found, let alone a distinct economy in the sense denoted, by the early nineteenth century.

Keith Tribe's *Genealogies of Capitalism* comes closest to my position. Tribe poses the central question: "In precisely what form does 'the economy' irrupt into economic discourse?" (1981, 125). If I read him correctly, he sees Ricardo as the critical figure in that story, thus casting doubt on the long-standing assumption that the Enlightenment political economists had in mind the same image of the world as subsequent economists. In another work, *Land, Labour and Economic Discourse,* Tribe argues that, for eighteenth-century political economists, especially the Physiocrats, nature was the prime mover of economic processes. Man was simply the "midwife" (Tribe 1978, 91, 95). I concur with the main arguments in both books but believe that they demand qualification and require bolstering with more evidential support. For one thing, they need to be joined with scholarship in the history of the natural sciences, something that Tribe does not enlist, and, for another, these theses would benefit from an understanding of the broader context of the secularization of science.

Other scholars have written on the connection between political economy and nature, notably Charles Clark in his *Economic Theory and Natural Philosophy* (1992). This book, while insightful, makes an argument very different from mine. It insists that "the aspiration [to explain economic events as natural phenomena] is uniform" from the seventeenth century up to the present (Clark 1992, 31). Clark mostly focuses on Smith and Mill, but he argues that, even for late-nineteenth-century economists such as Marshall, one finds the "Natural Law Outlook." The term *outlook* is Clark's own, and he defines it as an appeal by economists to the role of laws of nature both in the orderliness of the world and in moral amelioration. While I emphasize other facets of the intrusion of the natural into economic thinking, and while I agree with Clark that these concepts are present in the eighteenth century, I disagree with his interpretation of post-Ricardian economics. Mill, as we will see, did not believe in a world designed by God or in a morality grounded in the laws of nature. His thought constitutes a profound departure from the way in which economic phenomena were previously conceived and was of central importance in laying the conceptual foundations for the ascendant neoclassical theory.

Another related work to which I am much indebted is Vernard Foley's *Social Physics of Adam Smith* (1976). Foley delves into Quesnay's use of Cartesian vortices and Smith's infatuation with Epicurean atomism as well as the links between Smith and Newton. Deborah Redman's *Rise of Political Economy as a Science* (1997) covers a longer period, from the seventeenth century through the nineteenth, but the leitmotiv there is methodology. Again, there are many solid insights, but they pertain more to the adoption of hypothetico-deductive reasoning and other "scientific" methods than to the theoretical content itself. However, insofar as one cannot separate method from content, Redman's study is a valuable asset.

While Aristotle remains one of the earliest known writers on economic topics, the most fundamental economic assertion that he bequeathed to the scholarly world is to be found in his treatise *On the Heavens,* where he maintains that nature does nothing in vain.[22] In a sense, this is the most essentially economic claim that one can make: nature is fully efficient. There are no superfluous entities or processes. Aristotle sees this borne out in the wondrous symmetries of animals and plants and in the direct paths that physical bodies describe when in motion. Chapter 2 will trace the evolution of this concept among natural philosophers, noting that one of the first to develop it in some detail was Carl Linnaeus, the great Swedish naturalist of the mid-eighteenth century. Linnaeus, interestingly, also cultivated an account of economic development that was rooted in the German cameralist tradition. But, in many respects, he provides us with the first full-fledged description of an economy, only it is a description of an economy that encompasses everything, including plants, animals, and the deity. This helps underscore all the more the close links between economic thought and natural philosophy that prevailed in the eighteenth century.

Chapter 2 also examines the secularization of science. This is important if we are to come to terms with the equally difficult question of the source of conviction in the existence of the economic order. In the Middle Ages, and even in the early modern period, most prices were fixed by a local authority. The tradition of the just price linked it to the system of divine rule, but there was no obvious reason to suppose that prices were governed by natural laws. Eighteenth-century savants, by contrast, shared the belief that economic phenomena such as prices were grounded in a natural order. To make sense of this development, it is important to look also at appeals to God or a deity as the creator of the economic order. As Daston has observed: "Because God was 'the Author of nature,' the natural order was ipso facto a moral order" (Daston 1998, 157). The gradual denaturalization of the economic order is, thus, intimately linked to the ongoing secularization of science. A shift toward treating economic phenomena in terms of human agency also diminished the efficacy of the deity.

Chapter 3 examines the contributions of the French thinkers, notably Quesnay and Turgot, both of whom drew direct linkages between natural and economic phenomena. Indeed, the point has been made so often that it hardly needs further elaboration. I also survey, albeit briefly, some of the subsequent developments in French economics. But there is really no one figure in nineteenth-century France who had the same impact as Ricardo, Mill, or Marshall. Jean-Baptiste Say's works were widely read but perceived, incorrectly, as distillations of Adam Smith. The two best candidates for original and influential contributions are A. A. Cournot and Léon Walras, but both failed to gain an audience among economists until the end of their careers and were fully appreciated only in the early to mid-twentieth century.

In chapters 4 and 5, I examine the two most prominent Scottish political economists, Hume and Smith. My primary emphasis is on their efforts to join economics to nature. Numerous scholars have emphasized the fundamental importance of natural science in the development of moral philosophy among the Scottish Enlightenment writers. Roger Emerson, for example, takes the emphasis on natural science in the first edition of the *Encyclopaedia Britannica* (an Edinburgh publication) as highly indicative of "how thoroughly the great generation of Scottish thinkers had made science and its methods part of the intellectual culture of their time" (Emerson 1990, 25). He points out that most of the leading professors at the Scottish universities who taught moral philosophy were also versed in natural science. Inspiration from Newton, Boyle, and the early-eighteenth-century mathematicians was garnered, not just in the form of method, but in terms of metaphysical content as well. Scottish Enlightenment figures who were "thinking about *substances, causality, purpose, life, power,* or *agency* were also thinking about the metaphysical bases of morals" (Emerson 1990, 34).

With chapter 6, I travel south, as the clock ticks, to the work of the great English economists, Malthus, Ricardo, Senior, and McCulloch. I argue that there was a gradual shift toward the role of human institutions as the point of origin for economic phenomena and that commitment to providing a detailed account of human nature and of wealth as a natural phenomena had waned considerably. This was also the era in which the concept of an economy became explicit in terms of what we would now call *macroeconomic relationships.*

Chapter 7 takes up Mill's conception of nature and its implications for his political economy. I argue that a radical reconfiguration transpired, one that in many respects laid the groundwork for what came to pass in the 1870s with the efforts of the early marginalists. I also examine the importance of new movements in psychology, or the science of the mind as it was then known, and the Darwinian revolution. Finally, chapter 8 offers some evaluations of the early neoclassical period, particularly in the case of the English economists, Jevons, Edgeworth, Wicksteed, and Marshall, and argues that they brought about a significant rupture with the past. Despite numerous changes and developments in economic theory—game theory or the new institutionalism, not to mention considerable mathematical refinement—we have lived with their vision ever since.

CHAPTER TWO

Related Themes in the Natural Sciences

A purpose, an intention, or design strikes everywhere the most careless, the most stupid thinker; and no man can be so hardened in absurd systems, as at all times to reject it. That nature does nothing in vain, *is a maxim established in all the schools, merely from the contemplation of the works of nature, without any religious purpose.*

—David Hume, *Dialogues concerning Natural Religion*

NATURE DOES NOTHING IN VAIN

The ancient Greeks bequeathed to us the idea of a balance in nature, to whit that nature has within it a tendency to preserve its structural identity and sustain its original plan (Lloyd 1987, 319–36). For Plato and Aristotle, the world of appearance was always in flux, always in a state of becoming rather than of being. Nevertheless, that which underlies the given of experience is constant and, thus, induces a balance. Plato describes in the *Timaeus* an ingenious version of atomism, whereby the four basic elements are reducible to two types of triangles. Aristotle unleashed the doctrine of essentialism, which was of much greater utility to subsequent scientific inquiry than Plato's theory of the forms (see Lennox 2001, chaps. 7–8). But both of them subscribed to a doctrine of elementary substances—earth, air, fire, and water. Correlative to a belief in the balance in nature was a belief in the conservation of matter. Whatever the account given of the creation of the world, not a single particle had been added or lost since the beginning of time. Moreover, everything had its rightful place, and, thus, everything was true and just in and of itself (see Lloyd 1970, chaps. 6–8).

One of the more striking properties imputed to nature was efficiency. Nature, it was maintained, does nothing in vain. There is nothing superfluous or without a purpose. Aristotle asserted this principle with respect to celestial bodies, but he extended it to his studies on animals. Nature does what is best for each animal; if there is a better configuration, that is the one nature takes. This principle is manifest in the distribution of teeth and tusks among assorted quadrupeds, in the presence of fins rather than limbs in fish, and in the many stingers and spurs that nature gives animals for attack and defense (see Lennox 2001, chap. 9).

In the early modern period, the majority of natural philosophers embraced both the principle of a balance in nature and the idea of nature's efficiency—but in conjunction with the idea that the world had been designed by the Christian God. Each creature had its place in the grand scheme of things, and each part of every plant and animal was designed for a purpose, to sustain this balance. Everything was, thus, fully efficient; there was no waste in nature. These principles motivated William Harvey's theory of the circulation of the blood, which made sense of the large ventricles of the heart and the valves in the veins. As John Ray observed in his 1691 tract, *The Wisdom of God Manifested in the Work of the Creation,* every creature and plant was endowed with the means to further its survival. Where they had lost the ability to defend themselves, as in the case of sheep, they had been placed under man's stewardship to ensure their continuation.

Natural philosophers of the early modern period also devised a mechanical philosophy, although one with strong theological undertones (see Westfall 1958; and Brooke 1991). Matter may be reducible to microscopic atoms or corpuscles, but the deity was the original, if not the ever present, source of motion. Galileo, Descartes, Gassendi, Boyle, and Newton were leading exponents of the mechanical philosophy, although each one had a different construal of the nature of the atoms, the possibility of a vacuum, and the role of God in sustaining motion. Where they converged was in a belief in the uniformity of nature. For reasons of simplicity and elegance, God had created a world of only a small number of different-sized particles and a limited set of forces. Newton, for example, recognized the possibility of several kinds of force—gravity, fermentation, cohesion, and electric and magnetic attractions—but all, he speculated, were subject to the "general Laws of Nature" (1704/1979, 401). In short, nature was orderly. We may never comprehend more than a superficial portion of God's sensorium, yet the evident uniformity of the heavenly bodies and the symmetry of animal bodies speak overwhelmingly of a world subject to law.

Enlightenment philosophers undertook a fuller development of Newton's philosophy of nature. In Scotland, some of the leading proselytizers were David Gregory, John Keill, and Colin Maclaurin, while, to the south, Richard Bentley and Roger Cotes ensured that Newton was taught at Cambridge. On the Continent, Newtonianism was challenged by the legacy of Descartes and Leibniz but spread nonetheless, thanks to the teachings of Hermann Boerhaave and the popular accounts by Francesco Algarotti and Emilie du Châtelet. Newton's mathematical and mechanical contributions were greatly extended and refined by Maupertuis, Clairaut, and Euler, to name but a few. By the end of the eighteenth century, physicists such as Lagrange and Laplace had devised a rigorous theory of the material world, notwithstanding some fundamental unsolved problems such as a complete mathematical solution to the three-body problem. Bound by the principle of least action and that of the conservation of momentum, the motion of material bodies could be elegantly depicted by a set of simultaneous equations. Although Newton's metaphysics received considerable criticism from the Cartesian and Leibnizian philosophers on the Continent, his doctrine of gravitational attraction and mechanical forces in general became the orthodoxy following some definitive confirmations. An expedition to measure the shape of the earth in 1739, the mathematical solution of the moon's orbit by Alexis Clairault in 1747, and the predicted return of Halley's comet in 1759 put to rest most remaining skeptics (see Gascoigne 2003).

By the 1720s, Newton's physics faced little opposition among British mathematicians and was entrenched in the university curriculum of both the English and the Scottish universities. Colin Maclaurin, who was the "life and soul of the University of Edinburgh" during his term there as professor of mathematics (1726–46), also took to heart Newton's plea to move beyond natural philosophy to the more difficult questions of moral philosophy and, thus, to serve God better (see Lawrence 1982, 3). His *Account of Sir Isaac Newton's Philosophy* (1748) was the most influential text on the subject in Britain at the time. A broader theological reach was also part of the teachings of Bentley, Samuel Clarke, and Robert Greene (see Gascoigne 2003).

It is important to grasp that mathematics in that period included much of what we would consider physics today and even other subjects that are outside the natural sciences. Maclaurin, for example, taught astronomy, optics, mechanics, gunnery, geography, fortification, and surveying as well as geometry, algebra, and fluxions (the Newtonian version of what became the calculus).

Much of this went under the rubric *mixed mathematics,* in contrast to experimental physics. This latter field again ranged over areas that one might suppose belonged elsewhere: electricity and magnetism, airs and fire, but also human physiology. Many of the great physicians of the age were also physicists; indeed, the terms *physician* and *physicist* were used interchangeably. Hermann Boerhaave, Archibald Pitcairne, and William Cullen were salient examples of this conflation, but there were countless more (see R. Porter 1995). Efforts had been made since Descartes and Borelli to devise a mechanical study of human physiology, or *iatromechanics* as it was called (see Hankins 1985, 114–15). It was not until around 1800 that physics fully jettisoned investigations that would henceforth belong rather to medicine, biology, and chemistry (Heilbron 1979, 362); given current hybrids (e.g., biochemistry), such purity was not long-lived.

Most of the work on pure and mixed mathematics was carried out by Continental figures during the Enlightenment, with Leonhard Euler (1707–83) the most celebrated of all (see Bos 1980). Other major figures were Jean D'Alembert, the Bernoullis (Daniel, Johann, and Jakob), Joseph-Louis Lagrange, Pierre-Simon Laplace, and Pierre-Louis Maupertuis. Their contributions to mathematics and rational mechanics were eventually absorbed by economic theorists, toward the end of the nineteenth century and the first half of the twentieth (see Mirowski 1989; and Ingrao and Israel 1990). Earlier efforts to wed mathematics and economics—notably by William Whewell and A. A. Cournot—did not spark a movement at the time, Whewell's and Cournot's fame as philosophers notwithstanding. Although mathematical physics set the epistemological standard to which most other disciplines aspired, for the period under consideration here it was essentially experimental physics and natural history and not mathematical physics that seeped into the content of classical economics.[1]

EXPERIMENTAL PHYSICS

Historians of science have established that two distinct research traditions followed in the wake of Newton's achievements. One continued the project of mathematical mechanics, as laid out in the *Principia* (1687); the other cultivated the experimental sciences as described in the *Opticks* (1704).[2] For those like Berkeley and Hume, who found the mechanical philosophy unsustainable, the experimental tradition was the more plausible direction in which to take science (see Schofield 1970). This latter tradition focused on the study of the aether and its various manifestations—heat, light, electricity, and magnetism. By the 1730s, the study of these mysterious entities was unified by the doctrine of subtle or imponderable fluids, much as a century later they would be brought under the rubric *energy*. These substances were described as imponderable (having no weight) and were, thus, as Newton had recognized, inexplicable in terms of the mechanical philosophy.

London and Leiden took the lead as important centers for experimental science in the first half of the eighteenth century. Francis Hauksbee, curator of instruments at the Royal Society under Newton's reign, was without parallel as a designer and builder of new instruments. His most famous invention was the electrostatic generator, first built in 1705. By rubbing a rotating barometer (evacuated glass globe), one could draw off a series of electric sparks. By the 1730s, Stephen Gray and Charles-François Dufay had devised the means to conduct the electric fluid along silk threads, and this opened the way to a much wider range of experiments.

Two Dutch experimentalists, W. J. 'sGravesande and Pieter van Musschenbroek, made their way to London in the 1710s and imbibed a strong tonic of Newtonianism. They subsequently turned the University of Leiden into the most important center for experimental philosophy, particularly the study of electricity (see Ruestow 1973). The understanding of electricity was much facilitated in 1746 with Musschenbroek's construction of the first electric condenser, appropriately known as the *Leyden jar.* Now the electric fluid could be stored in a jar and extracted at a future time when an experiment warranted its use.

In his Query 31, Newton had suggested: "Bodies act one upon another by the Attractions of Gravity, Magnetism and Electricity" (1704/1979, 376). But he had also acknowledged that little was known about the nature of these forces. Magnetism's nature became less enigmatic once magnetism was linked to electricity by Hans Christian Oersted and Michael Faraday in the 1820s and 1830s, but until the 1920s, magnetism was less well understood than electricity. New apparatus devised in the early eighteenth century proved a major impetus to electrical inquiry. Hauksbee's generator and the experimental innovations of Gray, Dufay, and Musschenbroek made significant inroads toward a qualitative understanding of the electric fluid. Benjamin Franklin was celebrated for showing that lightning was composed of it. His experiments on electricity were circulated in 1747 in the form of letters to Peter Collinson that subsequently appeared in print in the *Philosophical Transactions of the Royal Society* of 1751. According to I. Bernard Cohen, Franklin's correspondence with Collinson was one of the most widely read scientific documents of the Enlightenment and inspired investigators across the face of Europe (see Cohen 1990, 28).[3]

Once electricity could be continuously generated by a machine, conducted along silk threads, and stored in a condenser, the learned world became all abuzz with excitement at this versatile fluid. Historians of science describe the rage with which electrical parlor games and laboratory displays swept Europe during the 1740s, superseding even the popular game of quadrille (see Heilbron 1979, 369; and Hankins 1985, 55). The theatrical spin-offs of electrical experimentation were numerous and varied. Stephen Gray suspended a young boy by silk cords and showed how his electrified state could attract small bits of paper. The Abbé Nollet was famous for having electrified 180 gendarmes for the amusement of the French royalty. As Geoffrey Sutton has argued, these displays primarily served the purpose of "a decorous public exercise, a performance, a demonstration of what was known rather than an investigation into what was not" (Sutton 1995, 288). Electrical experiments were also often described in the popular learned periodicals of the time, such as the *Gentleman's Magazine* (see Schaffer 1983).

Electric shock treatments in medicine commenced soon after the construction of the Leyden jar in 1746, prompting numerous controversies over their efficacy, particularly in the years 1749–52. One prolonged debate ensued between William Watson, the leading electrician at the Royal Society at the time, and Nollet (Schaffer 1983, 12–14). Franklin also experimented with electric shock treatment for paralysis (Cohen 1990, 55). In the case of lightning rod experiments, there were sensationalist, death-defying acts of bravery, one of which resulted in the inglorious end in 1753 of the Prussian physicist Georg Wilhelm Richmann.

Electricity was easily conceptualized as a fluid of subtle particles. It could flow from one body to the next, but it could also be retained in a vial (the Leyden jar) and extracted over the course of several days until it dissipated into the surrounding air. Franklin conjectured in his Collinson letters that the electric fluid was common to all bodies and that "common matter is a kind of sponge to the electrical fluid." Bodies become charged when additional amounts of the fluid are poured into it. "In common matter there is (generally) as much of the electrical as it will contain within its substance. If more is added, it lies without upon the surface, and forms what we call an electrical atmosphere; and then the body is said to be electrified" (Franklin 1754/1968, 315). Similarly, if a body is charged by another, the principle of conservation means that the fluid is deficient in the donor body and excessive in the recipient one. The electric potential, as we would now call it, could be greater or less, depending on the work done to move the fluid from one place to another, and, the larger the potential, the greater the shock experienced. Franklin emphasized the restoration of an equilibrium once the bodies were connected and, thus, inferred that the electric fluid obeyed the principle of conservation. Furthermore, different substances had different capacities to hold the fluid, just as a sponge could hold more water than a rock. Much of the debate at the time focused on whether there was one electric fluid or two, to take account of the attractive and repulsive forces. But almost no one challenged the idea of a fluid per se; the properties of diffusion, conservation, and capacity were all too manifest.

Electricity also held out the hope of unlocking the secrets of corpuscular attraction and repulsion and possibly the nature of life itself. Already by the 1730s, physicists had come to the realization that virtually every substance could be electrified and, thus, speculated on the ubiquity of the electric fluid. As a subtle fluid, the electric material could flow and be held in place (hence the concepts of conductivity and capacity) and, it was assumed, filled the interstices between material particles, much as sand fills the spaces between glass marbles. Some natural philosophers began to conceive of the electric fluid as the most essential substance, that which bound all the universe together in the sense of an aether. Boerhaave, Cullen, and Hutton were all of the view that fire, heat, and electricity were modifications of the same substance, the aether that Newton had cited (Lawrence 1982, 3).

Most historians have portrayed eighteenth-century British natural philosophy as a shadow of its former self. Haunted by Newton's ghost, and impervious to the more sophisticated mathematical techniques of the Continental savants, few British natural philosophers, if any, could hold a torch to a Boyle or a Hooke, let alone Newton. But British natural philosophy of the period from the 1740s to the 1780s was replete with investigations into electricity and the subtle fluid theory more extensively. Such prominent naturalists as John Pringle, Stephen Hales, and John Hunter inspired investigations into the medical application of fluid theories. Robert Schofield refers to a long list of "physicists" who published books or articles on the subject— Henry Miles, John Freke, Benjamin Martin, William Watson, Benjamin Wilson, Bryan Robinson, Richard Lovett, John Ellicott, and Robert Symmer. Moreover, many saw this as part of Newton's legacy. Wilson, for example, published in 1746 *An Essay towards an Explication of the Phenomena of Electricity, Deduced from the Aether of Sir Isaac Newton.* Two years later, Ellicott published several influential "Essays towards discovering the laws of Electricity" in the *Philosophical Transactions.* To this list one can add the pathbreaking late-eighteenth-century experiments of Joseph Priestley and Henry Cavendish.

Electricity was in many respects the paradigmatic field of experimental science in the British Enlightenment. But it was part of a broader theoretical investigation into subtle fluids, one that had been initiated by Stephen Hales in his experiments with the absorption of air and liquid by plants (reported in the 1727 *Vegetable Staticks*). William Cullen, Joseph Black, and James Hutton, who dominated the scientific community of Scotland for about fifty years, were enthusiastic about the concept of a subtle fluid, particularly as it applied to the phenomenon of heat. Cullen worked intensively on the problem of evaporation and also lectured on the aether and its relation to heat and electricity in his courses at the University of Glasgow. Black was much indebted to Franklin's analysis of the electric fluid for his own celebrated studies on latent heat. These insights in turn inspired James Watt in his construction of the separate condenser. The concept of the subtle fluid thus had engineering applications as well as medical ones. Hutton used the fluid of heat as a major force in his account of geologic transformations. As we will see in subsequent chapters, certain properties of these investigations filtered into the central ideas of the leading economists at the time, notably, Hume, Smith, Quesnay, and Turgot. Concepts of nature were, thus, central to classical political economy.

NATURAL HISTORY AND THE OECONOMY OF NATURE

The Aristotelian notion of purpose and efficiency in the world of plants and animals was greatly revised and expanded by Carl Linnaeus in his popular tract *Oeconomia naturae* (1749).[4] Drawing specifically on the detailed accounts of adaptation in the writings of John Ray and William Derham, Linnaeus argued that the world of plants and animals was so perfectly arranged that each species provided just the right amount of nourishment to others in the *scala natura*. The rate of propagation and the longevity of individuals had been consummately calculated by the Creator to sustain this balance. And if, by some natural hazard such as a storm or an earthquake, the balance was disturbed, certain creatures, insects in particular, could expand their numbers rapidly so as to restore the balance. For Linnaeus, insects were literally the "police" of nature, the troops that bring calm and order to a troubled region.

Linnaeus did much more than provide us with a workable system of taxonomy, albeit one that was arbitrary and artificial. As Lisbet Koerner (née Rausing) has argued, he was also an economist with strong cameralist leanings (Koerner 1999). He devised a grand scheme for an autarkic Sweden, advocating that tropical fruits and vegetables be domestically cultivated in greenhouses. He even tried, although with limited success, to grow tea and cocoa in southern Sweden, hoping that this would bring the whole country economic independence. Linnaeus paradoxically put great stock both in the fixity of the species and in the adaptability of life-forms to new regions, given human artifice.

Here is the first picture of an economy, that is, of a complex set of relations sewn together by supply and demand, with substitutability and mobility of resources. Man's activities are seamlessly joined to those of plants and animals, even to the earth's crust and atmosphere. As Linnaeus observed in the 1739 *Tal om märkvärdigheter uti insecterna:* "Everything arranged by the omnipotent Creator on our globe is performed in such a wonderful order that there is not one thing that is not dependent for its existence on the support of another . . . The earth becomes the food of the plant, the plant that of the worm, the worm that of the bird and the bird often that of the beast of prey. . . . Man who turns everything to his needs, often becomes the food of the beast or bird or fish of prey or of the worm and the earth. So all things go round" (quoted in Lepenies 1982, 20–21).

In another work, the posthumously released *Nemesis divina,* Linnaeus weaves in the role of God to create a divine or moral economy that supervenes on the worldly one.[5] Everything is part of some scheme of cosmic justice; there is retribution and revenge for each incursion on the balance of nature. Offenses against the laws of nature are punishable for Linnaeus. As Koerner put it in regard to the *Nemesis divina:* "Linnaeus never distinguished the moral from the natural" (Koerner 1999, 74).

At the time of Linnaeus, the word *oeconomie* was mostly used in the Aristotelian sense of household management. But it had also acquired analogues at the level of the state, particularly in the sense of the flow of commerce. Linnaeus read both meanings into nature and fashioned an economics of the nation-state out of his economy of nature. God's great household evinced his unbounded wisdom in arranging the kingdoms of plants and animals with such efficiency. Each species reproduced in just the right numbers to provide sufficient nourishment for others while allowing them to sustain themselves as well. The Creator also put a limit to the appetite of each and adapted each creature to its own special habitat, whether it be the bat to its cave or the camel to its desert (Linnaeus 1791/1977a, 95).

An equilibrium was manifest both in the number of offspring per species and in the average life span of each. There was, simply put, no waste. All that was produced was consumed. God had seen to the provision of scavengers to rid the earth of its decaying corpses. Even the seemingly random distribution of seeds, via the digestive tracts of birds and other winged devices, enabled supply to equal demand: "By the Oeconomy of Nature, we understand the all-wise disposition of the Creator in relation to natural things, by which they are fitted to produce general ends, and reciprocal uses. . . . Whoever duly turns his attention to the things on this our terraqueous globe, must necessarily confess, that they are so connected, so chained together, that they all aim at the same end, and to this end a vast number of intermediate ends are subservient" (Linnaeus 1791/1977a, 39–40).

Linnaeus's *Oeconomy of Nature* offers numerous instances where economic categories and phenomena are imputed to the more "excellent oeconomy of nature" (Linnaeus 1791/1977a, 126). In the later *Polity of Nature* (1760), Linnaeus depicted the mosses, grasses, herbs, and trees as the peasants, yeoman, gentry, and nobility, respectively. Insects were relegated to police nature and, thus, to maintain law and order: "Thus we see Nature resemble a well regulated state in which every individual has his proper employment and subsistence, and a proper gradation of offices and officers is appointed to correct and restrain every detrimental excess" (cited in Stauffer 1960, 240).

The oeconomy of nature was conceived as one that was completely full of life. At creation, and, of course, immediately after the Flood, there were only pairs of creatures. But it did not take long, given the geometric rate of reproduction, for species to spread throughout the globe: "The great Author and Parent of all things decreed, that the whole earth should be covered with plants, and that no place should be void" (Linnaeus 1791/1977a, 67).[6] Linnaeus considered the possibility of the globe's population doubling or tripling and concluded that, if it did, humanity would perish since "the surface of the earth can support only a certain number of inhabitants" (119). He thus depicted the economy as full and stationary. He also assimilated what we would now call the *hydrologic cycle* into his economy of nature. Water, air, and the earth's crust were all part of the continuous exchange of life-forms.

Linnaeus was also instrumental in establishing economics as a science. As a founding member of the Swedish Academy of Science, he made one of its primary objectives the promotion of economic inquiry. As he stated in the first volume of the academy's *Acts* (1740): "No science in the world is more elevated, more necessary, and more useful than Economics, since all people's material well-being is based on it" (quoted in Koerner 1999, 103). He also oversaw the establishment of chairs of economics at every Swedish university but one and instructed his advanced students to travel to the far reaches of the globe so as to further the economic enrichment of Sweden. For Linnaeus, all plants were potentially adaptable to the Swedish climate, given enough time and the cultivation of hardier strains. Expanding the Swedish stock of cultivars in turn would make Sweden economically self-sufficient. The protectionist and parochial vision of cameralism was, thus, given a solid grounding in biological nature.[7] According to Koerner, for Linnaeus "an economist without knowledge about nature is therefore like a physicist without knowledge of mathematics" (1999, 103).

Although Linnaeus was unwavering in his belief in God, the Enlightenment witnessed an increasing number of secular philosophers, notably Hume and Voltaire. The doctrine of spontaneous generation garnered support in the 1740s and, thus, lent weight to a thoroughgoing materialism. Life might be simply the product of a chemical or an electrical event in some primordial sludge. Investigations into the history of the earth had significantly challenged the biblical account and had, thus, opened the door to evolutionary doctrines in the hands of Georges-Louis Buffon, Jean-Baptiste Lamarck, and Erasmus Darwin. As Hutton was to conclude in his *Theory of the Earth* (1795), geologic strata and the ongoing motion of the water and volcanic forces implied "no vestige of a beginning, nor prospect of an end" (cited in Hankins 1985, 155).

Charles Lyell, the leading geologist in the early Victorian era, had read Linnaeus carefully but was equipped with both the concept of extinction, as devised by Georges Cuvier ca. 1800, and an immense geologic scale, from Hutton among others (see Gillispie 1959; Rudwick 1976; and Moore 1986). In order to explain the existence of great organic diversity and the seeming absence of natural monopolies among species, Lyell postulated the creation of new species to fill the gaps caused by extinction and the ever-changing surface of the earth: "We could imagine the successive creation of species to constitute, like their gradual extinction, a regular part of the economy of nature" (Lyell 1830–33, 2:179). He did not dwell on the means by which new species were produced. This was, in short, that "mystery of mysteries" that Charles Darwin set out to solve.

According to some historians, the Linnaean concept of the economy of nature remained more or less intact in Lyell and Darwin until it was superseded, in 1866, by the separate pursuit of what was called *oecologie* by Ernst Haeckel.[8] Others, however, have argued that one of the important turning points in Darwin's conceptual genesis was his breach with the Linnaean economy, stimulated by his reading of Malthus on population and Henri Milne-Edwards on the division of labor. The latter had acquired the principle from Jean-Baptiste Say, so the economics was absorbed secondhand (see Limoges 1970; and Schweber 1977). In my view, however, Lyell was as pivotal a thinker on this subject (see Schabas 1990a). It was his conception of the economy of nature, as articulated in his three-volume *Principles of Geology* (1830–33), that reflected new developments in political economy. If Darwin's economy of nature appears to be "just a vast generalization of Ricardian economics" (Hardin 1960, 1295), it is because he learned his economics in a roundabout fashion from Lyell.[9] There is much to support Darwin's remark: "I always feel as if my books came half out of Lyell's brain" (quoted in Rudwick 1971, 209).

Lyell observed that the supply of food equals the demand among plants and animals and that there are forces to restore this balance in the case of geologic and biological disruptions. He depicted the natural realm as in a state of dynamic equilibrium, in which reproductive rates and life spans of individuals, as well as the rate of extinction and creation of new species, adjust to one another so as to bring about a "perfect harmony" (Lyell 1830–33, 2:42). He believed that there were different rates of stabilization, depending on the kind of disturbance. In the case of a temporary climate change, instability was short-lived, but, in the case of the arrival of a new plant or animal, it might "require ages before such a new adjustment of the relative forces of so many conflicting agents can be definitively settled" (145). Human beings, "considered merely as consumers of a certain quantity of organic matter," are no different from any other species in this respect (146). Lyell considers the possibility of man having increased the total yield of a given region. As we cultivate land, it is difficult to suppose that "we have not empowered it to support a larger quantity of organic life" (147). But, he insists, our population has grown only at the expense of other life-forms. Lyell's economy of nature was in a steady state.

With characteristic perspicacity, Darwin grasped that diversification brought about an expanding economy of nature. One of his central principles is "that the greatest amount of life can be supported by great diversification of structure" (Darwin 1859/1968, 114). In one experiment, he weighed the grasses grown on two identical plots of land, one sown with just one type of grass, the other with several distinct genera. The second plot yielded a far greater "biomass" than the first. Hence, he observed: "In the general economy of any land, the more widely and perfectly the animals and plants are diversified for different habits of life, so will a greater number of individuals be capable of there supporting themselves" (115). Nature is no longer full, nor is it in a state of equilibrium. Insofar as "natural selection is daily and hourly scrutinising," life-forms are in continual flux (84). Darwin thus broke with the Lyellian model of a fixed quantity of life oscillating in various forms about an equilibrium point. Because of Darwin's commitment to increasing diversity of life-forms, the economy of nature has within it a propensity to grow and expand.

More significantly, Darwin's recognition that diversification increases the quantity of life mirrors the Smithian insight that the division of labor between trades is a function of the size of the market or the size of the economy more generally (see Schweber 1977). In the Scottish highlands, everyone must be a butcher, a baker, and a brewer, but, in the bustling city of London, one would find specialists by trade, even subtrades like pastry chef. Such diversification was possible only in a large market; the two grew in tandem. Similarly, the rise in the number of distinct species, say, in a plot of grass, was directly correlated with the growth in the quantity of life itself.

The concept of the economy of nature evolved in the hands of Linnaeus, Lyell, and Darwin. Linnaeus's conceptual apparatus is, as one would expect, preclassical. It is a commonsense static economy, with elements of competition and implicit ratios of exchange, but with few of the mechanisms later associated with classical political economy. But, in the case of Lyell and Darwin, we find numerous insights that have analogues in classical political economy. A significant modification of the concept of the economy of nature came with Lyell's recognition of time-consuming equilibrating processes, a feature that resembles Ricardo's theory of production and distribution. For Ricardo, a growth in population lowers the real wage, thus engendering a growth in production. This creates a differential profit rate, owing to different capital-labor ratios in different sectors of the economy, which results in a shift in resources into the more profitable sectors. The increase in supply in those sectors lowers market prices, until the profit rates in the labor-intensive industries are once again equal to the profit rates in the capital-intensive ones. Lyell depicts an economy of nature with similar equilibrating processes, ones that revolve around the difference in the food and physical conditions, or factors of production, required by the different life-forms. There is also evidence to support the view that Lyell absorbed this directly from his study of political economy. Martin Rudwick has noted of Lyell that he "saw his geology as a much more open-ended enterprise . . . [with] little sense of cognitive compartments and boundaries. . . . Lyell was able to quarry creative analogies from a wide range of extra-geological sources" (Rudwick 1979, 79).[10]

While Ricardo envisioned a growing economy, he worried more about the Malthusian problem of a stationary state, one where food supplies could not match population growth. To add to the scenario of doom and gloom, opportunities to invest in the capital stock were bound to diminish. Even the windfalls that came with machinery were for Ricardo cause for concern insofar as they could permanently displace labor. In this respect, Lyell was much closer to the Ricardian picture than was Darwin, who believed that the economy of nature could grow indefinitely. Lyell, we have seen, put much stock in the notion of a fixed quantity of life, which could be usurped from some species by others through an increase in relative population, but which could never increase overall. Darwin, on the other hand, had the remarkable insight that, through diversification, the human population could grow in conjunction with other species in the aggregate. In this sense, he emancipated the human economy from long-standing physical constraints in nature, in the same period during which John Stuart Mill, as we will see, furthered the denaturalization of the economic order.

For eighteenth-century savants, it was a given that man was an animal. Comparisons were rife—and not only to the more intelligent mammals. Animal imagery—such as that in Bernard Mandeville's *Fable of the Bees*—pervaded the period (see also Darnton 1985). Linnaeus, in creating the categories of mammals and primates, grouped us unequivocally with monkeys and apes (see Schiebinger 1993). Indeed, he believed accounts from faraway lands that some hirsute peoples formed a continuum with the apes and that some apes even played backgammon (Koerner 1999, 87). He also maintained, contra Descartes, that animals have souls. Hume, while sidestepping the spiritual question, argued that animals reason inductively from experience in the same manner as humans (see Pitson 1993). Smith, Rousseau, Montesquieu, La Mettrie, and Lord Monboddo drew numerous comparisons between the human and the animal realms, suggesting that the links were greater than had hitherto been grasped. The frequent reference to and depiction of animal oeconomies among eighteenth-century savants did no disservice to the concept of the human oeconomy. The two were seamlessly joined, as the work of Linnaeus best illustrates. Enlightenment philosophers saw the world as one integrated but complex web that linked the earth's crust, the atmosphere, plants, animals, and humans. Only the specific role of the deity was thoroughly queried.

THE SECULARIZATION OF SCIENCE

Historians of science have long argued that, at least until the Enlightenment, Western science was the "handmaiden" of religion, insofar as it served the church's need to instill a reverence for God's handiwork (see Grant 1986). One finds this worshiping of God's first bible even in Galileo, who marveled at the cicada's ability to make sweet sounds by seemingly invisible means. Arguably, one of the closest unions of science and religious faith in the modern period was the pursuit of natural theology, as exemplified by William Paley's popular tract of the same name published in 1802. Appreciation of the adaptative features of life-forms and the rich diversity of God's creations inspired numerous

Victorian scientists—William Buckland, Adam Sedgwick, and Richard Owen, for example—to pursue pathbreaking research in natural history, all the time professing themselves humble servants of the Anglican Church.

As John Hedley Brooke has argued, there is no simple story regarding the disentanglement of science from religion. That the two were intimately linked in the Middle Ages no one disputes. And that they are only sporadically linked today is again a claim that appears, with some qualification, to withstand scrutiny.[11] The secularization of science is, however, more a tale of fits and starts than one of linearity. Moreover, there are several different ways in which one might conceptualize the process of secularization. One is the shift of the practice of science, from clergymen to laymen. Another is the frequency with which scientific theories make appeals to God's efficacy. A third is the sense in which scientific knowledge, especially claims that are devoid of theology, superseded religious doctrine.[12] There are, of course, many other ways to get a handle on secularization in the modern era, notably in the form of cultural practices. But these can also be very misleading. One expert, Owen Chadwick, has disavowed the ready correlation of secularization and nonecclesiastical customs and habits.[13] Chadwick urges that we dig more deeply into the manner and force by which we in the modern age, including our scientific research, have come to function without religion.

The first sense in which science underwent secularization, namely, the diminution in the number of religious functionaries, transpired, albeit unevenly, from the early modern period up to the present. Today, few clergy make contributions to the frontiers of scientific knowledge. Three hundred years ago, probably more than a third of active "scientists" in Europe were ordained in one church or another or held a degree in theology, while, six hundred years ago, ca. 1400, science was conducted almost exclusively under the auspices of the Catholic Church. But the path by which the white laboratory coat replaced the cleric's collar was more frequented in France than in Britain prior to 1900. In the ancien régime, numerous natural philosophers were men of the cloth, such as Jean-Antoine Nollet and Étienne Bonnot de Condillac. The Jesuits in particular were active contributors. But there were also prominent and vocal critics of the priesthood, such as Voltaire and Dénis Diderot, both of whom advocated a more natural or materialistic approach to religious sentiment. After the Revolution of 1789, with the numerous reforms of higher learning and the establishment of elite institutes such as the École Polytechnique, few French scientists had clerical training or affiliations.

In Britain, by contrast, numerous ministers of the Church of England were engaged in scientific research, particularly natural history, right up until the mid-nineteenth century. This can be attributed partly to the fact that both Oxford and Cambridge required religious training and partly, as Brooke has argued, to the fact that the British did not have to confront papal authority, the Inquisition, or the Jesuit order. As a result: "A critical mentality and faith in progress could thrive in England within piety" (Brooke 1991, 200). This also partly explains why natural theology flourished primarily in the Protestant regions of Europe.

It might appear self-evident that the religious content of science itself would diminish over time, as laypersons replaced those with clerical associations. Presumably, nonclergy have less of a commitment to seeing science in theistic terms and less reason to ensure that their scientific views accord with church doctrine. This "desacralization" of scientific theory, however, was hardly monotonic since, for many prominent scientists, it is still imperative to reconcile their findings with their personal theology. Whereas much of the groundbreaking work in energy physics of the middle of the nineteenth century was done in secular terms, by the end of the century many English physicists were drawn to the idea that the electromagnetic ether was the seat of God himself, the so-called unseen universe (see Heimann 1972; and Smith 1998). Prominent physicists such as Gabriel Stokes and Lord Kelvin attended séances and other spiritual events; Thompson and Tait's famous 1867 treatise on natural philosophy did much to restore a spiritual core to the otherwise mechanistic discipline. Similarly, while mid-eighteenth-century views on the process of generation were tainted with atheism, much of that was erased by Louis Pasteur's triumphant efforts to defeat claims about the spontaneous generation of microorganisms as part of his mission to instill Catholic piety.[14]

Chadwick (1975, 167) astutely observed that most scientists, even those like Thomas Huxley who publicly challenged a prominent bishop, have realized that "the advances of science could hardly touch God." In short, while the number of atheists among scientists has no doubt grown over the past couple of centuries, the majority of scientists are believers and have some affiliation with a temple of one kind or another. The real contest between science and religion has been played out over the Bible as an authoritative source about the universe. Protestant efforts to promote a more liberal reading of Scripture were, thus, part and parcel of the reawakening of scientific activity in the late Renaissance, the celebrated Merton thesis (see Merton 1938). A more concrete challenge was Galileo's brilliant interpretation of the miracle of Joshua in terms of the heliocentric theory. But nothing holds a candle to the problems posed by Darwin's theory of evolution and its overt contradiction of the Book of Genesis. Even if one were to take the seven days as each representing millions of years, as many modern creationists purport to do, the sequencing in the Bible bears no resemblance to the paleontological record. In the history of science, Darwin and his cohorts stand out as the missionaries for the secularization of science.[15]

It would be misleading, however, to suggest that science in the modern period eschewed all allegiance to spirituality. The Romantic ideas of the *Naturphilosophie* movement that governed the work of Goethe, Schelling, and Coleridge left their mark on many branches of science (see Levere 1981; and Richards 2002). Darwin's theory had cut the thread to a biblical account of the natural world, but many a scientist still clung to some form of spiritualism, including Alfred Russel Wallace, who simultaneously and independently worked out many of the key elements of the theory of evolution. Even in physics, the second law of thermodynamics and Heisenberg's uncertainty principle were used by some to instill spiritual sentiments (see Hiebert 1986).

The final sense in which science underwent secularization is in terms of its ascendancy over religious doctrine as a source of authority. Probably the most significant figure in this transformation was August Comte, the instigator of positivism. Pure scientific knowledge, especially physics and astronomy, had triumphed over theology and metaphysics. Indeed, proper science is grounded only in the given of experience. Once science was extended into everyday language and politics, superstition and conflict would purportedly vanish, all in the name of order and progress. And, true to form, Comte then devised an elaborate "Religion of Humanity," replete with priests, catechisms, hymns, and other trappings that resembled Catholic worshiping. The contemporary set of scientific academies, museums, and ceremonies such as the awarding of the Nobel Prizes might also rival Christianity as a formidable social institution.[16] Certainly, the only contemporary projects to rival the construction of the great cathedrals of Europe, in terms of time and money, are cyclotrons and spaceships.[17]

Comte, of course, did not spring from a vacuum. Positivism owed much to Voltaire's natural religion, to Hume's skeptical empiricism, not to mention Kant's efforts to drive a wedge between science and ethics and render the latter equally universal and rigorous (see Brooke 1991, 163, 182, 209). Much of positivism has also been linked to the cleansing of scientific and ordinary language, as first envisioned by Condillac and Condorcet in the late eighteenth century. Alfred Jules Ayer, in *Language, Truth and Logic* (1936), his popular account of logical positivism, captured perfectly these sentiments when he bluntly ruled religious claims out of court: "Since the religious utterances of the theist are not genuine propositions at all, they cannot stand in any logical relation to the propositions of science" (1936/1946, 117). Only propositions that are verifiable are permissible. Any claim that appeals to God is, thus, set apart from science.[18] Positivism, moreover, by viewing science as politically liberating, served to empower factual claims at the expense of the Bible. Reverence for science and its achievements, as articulated, for example, in Alexander Pope's famous paean to Newton, served to diminish the sense of awe and mystery that was exclusively under the purview of religious practice.[19]

The nineteenth century also brought a closer union of science and industry, although nothing on the scale of the past fifty years. Nevertheless, there was enough to instill a reverence for the powers of science to bring about "the relief of man's estate," as once envisioned by Francis Bacon (1605/1955, 193). Increasingly, scientists of the nineteenth century, such as Charles Babbage, were keen to see their theoretical investigations result in practical applications. This is not to suggest that such links were absent in the past; almost every century had its own lesser version of an Archimedes or a Leonardo da Vinci. But it was only with the Industrial Revolution that a scientific spirit began to pervade the realm of economic production, whether in agrarian experimentation or the production of textiles and metals. Standardization and precision in engineering were also increasingly valorized, in keeping with the rigor and purported objectivity of the scientific community.[20] This bootstrapping of science to technology did much to bring on a more secular age.

Political economy from the seventeenth century through the nineteenth was probably more secular than any other branch of science, although there were numerous appeals to the deity and even, in the case of Malthus, an explicit union of economic analysis to the problem of theodicy (see Waterman 1991). That economic inquiry for the most part would be relatively secular might come as no surprise, given that the secularization of the Western world is often attributed to the emergence of capitalism and its toleration of private vices in the name of public benefits. The pursuit of lucre, not to mention the practice of usury, was explicitly condemned by the Bible. It took considerable ideological effort in the late Renaissance, as Albert O. Hirschman has so aptly depicted, to find a harmonious union between commercial activities and Christian doctrine (Hirschman 1977). The science of these activities, namely, economics, could not help but be partially secular. Throw in liberalism and utilitarianism, and one has a potent brew for a thoroughgoing secularism.

This is, of course, all a bit too simplistic. For one thing, Locke rendered us economic creatures in an Edenic state of nature. He also entreated God to underwrite our doctrine of rights. In that sense, Christianity lies at the core of modern economic liberalism, especially the right to property. Quesnay and his disciples wove frequent appeals to God's natural order into their economic tracts and used these to justify policies on trade and taxation. Adam Smith also gave the deity a providential role in his account of commercial society (see Viner 1972; and Hill 2001). Even utilitarian moral theory received a Christian gloss in the hands of Paley and Malthus (see Waterman 1996). Their contemporaries Richard Whateley and Thomas Chalmers readily joined Christian theology with mainstream political economy and treated the secular economy as part and parcel of sacred commands (see Alborn 1990; and Waterman 1991).

What enabled classical political economy to undergo secularization of the three varieties outlined above was its close and intimate ties with natural science. The two became increasingly secular more or less in tandem. Yet there were still a number of active contributors with overt Christian alliances right up until the end of the nineteenth century. Although God rarely made an appearance in economic treatises, he was also not entirely absent. Charles Kingsley, Philip Henry Wicksteed, Henry George, Jevons, and Marshall all expressed religious sentiments of one strain or another. Francis Ysidro Edgeworth cultivated an interest in spiritualism and attended some of the same séances as Stokes and Kelvin (see Creedy 1986, 38–39). If God was dead for a Marx or a Veblen, it was not the case that most late-nineteenth-century economists had nailed the coffin shut.

CONCLUSION

This chapter has offered brief sketches of a number of key developments in eighteenth- and nineteenth-century science. First, we have looked at post-Newtonian contributions to experimental physics, particularly the doctrine of subtle fluids such as electricity. We will see this play a part in the economic ideas of the Physiocrats, Turgot, Hume, and Smith. This argument is buttressed by Simon Schaffer's (1990) detailed account of the influence of theories of electricity, pneumatics, and heat on the political and economic ideas of Joseph Priestley and Jeremy Bentham toward the end of the eighteenth century (Schaffer 1990). All this will lend weight to the view that concepts of nature were deeply entrenched in economic thought at the time.

We have also seen that Linnaeus was one of the major economists of the Enlightenment, not only because of his cameralist policies and success in establishing the science of economics in Swedish universities, but also because of his developed account of the economy of nature. More than anyone of his time, he saw the human realm as seamlessly joined to the oeconomy of nature, an exchange system among animals, plants, minerals, and the atmosphere. As

Lisbet Rausing has recently observed, Linnaeus shared with his French and English naturalists the agenda of utilizing natural history for commercial expansion. But he also "went beyond such general sentiments to a more developed theory of the ultimate sameness of economics and the natural sciences" (Rausing 2003, 184). Arguably, the conflation of natural and economic phenomena never again received such dedication and devotion.

The concept of the economy of nature was further developed by Lyell and Darwin, with an apparent absorption of ideas from Ricardo and Smith, respectively. Economic and biological ideas were closely linked during the first half of the nineteenth century but then began to diverge. When we look in chapters 7 and 8 at the purported efforts to bring Darwin into mainstream economics, we will see that those linkages were relatively superficial. By the mid- to late nineteenth century, economics was no longer joined to natural history.

Political economy was also at the vanguard of the secularization of science. This is significant for several reasons. In the period when science was still far from being secular, namely, the age of Newton, order was easily imputed to economic phenomena as the product of the "Author of Nature." As a more secular age came to prevail toward the latter half of the eighteenth century, it could no longer be readily assumed that God had created an economic order or even had direct control over motives and passions in the commercial sphere. Economic development has rightly been identified as a central factor in the secular movements of the modern era, and economic theorizing reflected that trend to its very core (see Hirschman 1977).

As we will see in chapters 6 and 7, the economic order had to be explained in terms of human agency. Only psychological uniformities could make sense of the rhythm of prices and markets. And, as it became increasingly commonplace to see human psychology as entirely secular in its formation and regularity, so too the economy was increasingly conceived as a product of human agency and human institutions. Ironically, twentieth-century economists lost sight altogether of the need to seek or justify the law-like behavior of the economy. Stefano Zemagni has noted, correctly I believe, that "the list of generally accepted economic laws seems to be shrinking." The term *law* now seems old-fashioned, precisely because contemporary economists are less inclined to believe "that the only route towards explanation and prediction is the one paved with laws, and laws as forceful as Newton's laws" (Zemagni 1989, 104). More often, they are content to build models than search for new laws (see Morgan and Morrison 1999). We will see in the concluding chapter that this practice is part and parcel of the denaturalization of the economic order.

CHAPTER THREE

French Economics
in the Enlightenment

*In nature everything is intertwined, everything runs through circular courses
which are interlaced with one another.*

—François Quesnay

Few scholars, if any, dispute the claim that France was the center of intellectual activity during the Enlightenment, even if an English philosopher—John Locke—and a German philosopher—Immanuel Kant—are often credited with bounding the period (roughly 1690–1790). Nor is the recognition of France's central role meant to rob other places of their rightful claim on learning during the eighteenth century. Certainly, in many fields of inquiry, there were prominent contributors to be found in Italy, Sweden, the German-speaking regions, the Netherlands, Britain, and even Russia (see Porter and Teich 1981). But, in France, one can find eminent scholars in virtually every branch of the arts and sciences. Of the seventy-odd scientific societies that were established before 1800, almost half were to be found there. Learned individuals, including those in the Americas and Asia, flocked to Paris to rub shoulders with Diderot and D'Alembert, the editors of the *Encyclopédie,* or to attend the salons and the opera.[1]

In many respects, most characteristic of the Enlightenment, particularly the French Enlightenment, were efforts in moral philosophy, or the *social sciences,* as they came to be known by the end of the eighteenth century (see Baker 1964). Some shining examples are those of Voltaire, Montesquieu, and Rousseau, figures who easily eclipse in fame and influence the leading contributors to natural philosophy in France at the time, D'Alembert, Maupertuis, and Buffon. More impressively, these men met and exchanged ideas with some regularity, inspiring and spurring one another on. Each was convinced that, if new insights and a new rigor could be brought to bear on moral and political issues, moral philosophy would be transformed just as in the seventeenth century the natural sciences had been. Each believed that, insofar as the moral order mirrored the natural order, the betterment of human society could be achieved only by bringing it into harmony with physical nature. While each *philosophe* was remarkably secular for the period and, thus, much less inclined to grapple with God's efficacy than Descartes or Malebranche had been, the majority were nonetheless all drawn to the idea of a providential order laid down by the deity (see Brooke 1991).

All the aforementioned also wrote on economics, although some, such as Voltaire and Montesquieu, treated the subject only in passing. Voltaire endorsed Mandeville's views on luxury, sentiments that were echoed by Montesquieu. In books 20–23 of his *L'esprit des lois* (1748), Montesquieu addressed the phenomena of money, commerce, and population. As with other features of the moral realm, these were reduced to physical causes, specifically the climate. As he observed: "Though commerce is subject to great revolutions, it can happen that certain physical causes such as the quality of the terrain or of the climate fix its nature forever" (Montesquieu 1748/1989, 354). Montesquieu also allowed that commercial laws and practices could improve manners and, above all, induce peace, so, in that respect, moral traits were grounded in nature as well (see Hirschman 1977). He readily ascribed industriousness to the lack of conveniences in northern climates and laziness to the ease of procuring necessities in southern climates. Even drunkenness obeyed a gradient, increasing as one moved from the respective polar regions toward the equator (Montesquieu 1748/1989, bk. 14; see also Chamley 1975).

Rousseau wrote more extensively on economic subjects, on economic inequality, household management, and the "économie rustique," or "the art of knowing all the useful and lucrative objects of the countryside" (Spary 2003, 21).[2] He also drew analogies between "economies" and the body politic: "Commerce, industry and agriculture are the mouth and stomach which prepare the common subsistence; the public finances are the blood that is discharged by a wise *economy,* performing the functions of the heart, in order to distribute nourishment and life throughout the body" (Rousseau 1987, 114).

Rousseau also grappled with the concept of nature's gifts, an endeavor that drew him in the last fifteen years of his life to the study of botany and the composition of his popular *Reveries of the Solitary Walker* and *Botanical Writings* (see Rousseau 1781–82/2000). He was critical of the artificial cultivation of exotic plants and even the domestication of fruit trees. Nevertheless, nature was a storehouse of plants for human consumption, and, the more that could be extracted and utilized, the better. As Emma Spary has shown, a number of French economists in the mid-eighteenth century, such as Charles Ferrère Du Tot or Jean-François Melon, envisioned the great potential of botany as a means of economic expansion, at both the local and the global levels. Because of the medicinal and culinary value of exotic plants and the potential for cultivation in France, botany became an enterprise of the state. Antoine de Jussieu's Jardin du Roi was the first place where coffee and pineapples were grown in France, setting a pattern that was repeated later in the century by Linnaeus in Uppsala and Joseph Banks at Kew Gardens (see Koerner 1999; and Drayton 2000). These figures' respective missions to move economic management from the household garden to state-run enterprises sustained the trope of *oeconomie* as the art of resource management grounded in a reverence for nature's prodigality (see Spary 2003, 21).

Voltaire, Montesquieu, and Rousseau were part and parcel of an emerging discourse on political economy in mid-eighteenth-century France, most notably in the form of two short-lived journals, the *Journal Oeconomique* (1751–72) and the *Journal du Commerce* (1759–62). But there was a tradition that went back over a century, first to Antoine de Monchrétien's *Traicté de l'oeconomie politique* (1615) and, more significantly, to Pierre de Boisguilbert (1646–1714) and Vincent de Gournay (1712–59). Boisguilbert was perhaps the first, in the early eighteenth century, to grant nature a directive role in bringing about peace and order (see Christensen 2003, 115). He advanced the concept *laissez faire* in the sense of a providential order that reflects the influence of the metaphysician Bernard de Fontenelle (see Spengler 1984; Faccarello 1999; and Christensen 2003). In the mid-1750s, Gournay and his circle adopted the term *laisser faire* with policy measures in mind, namely, the dismantling of archaic restrictions on the trade of grain.

John Law (1671–1729) and Richard Cantillon (d. 1734) have often been grouped in the French tradition even though Law was born in Scotland and Cantillon in Ireland. Law wrote mostly about money and finance and is infamous for wreaking havoc on the French banking system through his promotion of paper notes and the Mississippi Company, which resulted in a speculative frenzy in Paris (see Murphy 1997). He and Cantillon had formed a partnership to undertake land purchases in the Mississippi Valley region. Both made huge fortunes and succeeded in escaping France right before the bubble burst in 1720. Cantillon's one text, the *Essai sur la nature du commerce en général* (1755), while published posthumously, may have circulated as a manuscript in the 1730s and 1740s and left its mark on both British and French economic thought (see Murphy 1986; and Brewer 1992). Cantillon stressed the idea of an equilibrium and the general flow of goods and services from one sector to another. And, while he privileged agriculture, much of his focus was directed toward the role of money. According to Murphy (1986, 251), Cantillon had a strong predilection for quantifying and collecting data, following in the footsteps of his much-admired William Petty. But, because we know very little about Cantillon's early life and education, it is impossible to assess his command of natural philosophy. And, because his influence was belated and indirect, I will treat him here as something of an outlier.

PHYSIOCRACY

François Quesnay (1694–1774) was the most prominent figure of eighteenth-century French economics and also the one who most instantiates the links between nature and economic phenomena, both physical nature and human physiology.[3] Quesnay also inspired a school of thought, Physiocracy, which flourished in the third quarter of the eighteenth century. Most prominent among the Physiocrats were Victor Mirabeau (1715–89), Pierre Mercier de la Rivière (1719–1801), and Pierre-Samuel Du Pont de Nemours (1739–1817), but there were at least a dozen more who contributed writings on the subject.

The school commenced in 1757 with a historic encounter between Quesnay and Mirabeau, in which the latter was persuaded to renounce his own views and accept the validity of Quesnay's. Nicolas Baudeau coined the term *physiocracy* to denote the "rule of nature," both for the sake of legitimation and to reflect the numerous conceptual borrowings from natural science and natural imagery. Momentum gathered in the early 1760s, and efforts were made, with some success, to influence economic policies, moving them in the direction of a less-restricted interprovincial trade of grain with a *bon prix* (high price). The Physiocrats journal, the *Éphémérides du Citoyen,* gained a strong following until it was discontinued in 1772 owing to government intervention. By the late 1770s, the school no longer existed. Quesnay died in 1774, and the bread riots preceding the Revolution did nothing to win converts to his policy of the *bon prix*. Attacks on his policy of a single tax by such diverse figures as Voltaire, Galiani, and Bonnot de Mably also contributed to the swift demise of Physiocracy.[4]

Quesnay started life as a surgeon and was most inspired by the teachings of the medical school at the University of Leiden. His philosophical mentors were Hermann Boerhaave and Stephen Hales, but he was also influenced by the iatrophysical tradition of Giovanni Borelli and Descartes (see Christensen 1994; and Banzhaf 2000). Because he was responsible for a breakthrough in the process of bloodletting and achieved some success in curing smallpox, Quesnay was stationed at Versailles as *médécin consultant du roi.* He was also made a member of the Royal Academy and the Académie des Sciences, although he made little use of these affiliations. Commentators have portrayed Quesnay as more a practitioner blessed with brilliant insights than a full-fledged scholar. Nevertheless, his system drew directly on concepts in mechanical philosophy and human physiology, thus suggesting a reasonable command of natural philosophy (see Banzhaf 2000).

Quesnay's early writings, notably his *Essai phisique sur l'oeconomie animale* (1736), explicitly addressed the "Author of Nature" and his set of universal laws. Quesnay accepted the central elements of Cartesian mechanics but found the ether problematic. Ether was thought to be made of microscopic particles, but, because it was the source of vitality for all animal bodies, it could not be the seat of its own motion. Malebranche's doctrine of occasionalism, which gives God a central and ubiquitous role in the continuation of mechanical action, thus appealed to Quesnay. Mechanical causes are consistent with the will of the prime mover, who is the true and proximate cause of all motion and order, both physical and moral. Thus, as Ronald Meek relates, Quesnay and his colleagues sought economic laws "which operated independently of the will of man and which were discoverable by the light of reason. These laws governed the shape and movement of the economic order and therefore (on the Physiocrats' basically materialist hypothesis) the shape and movement of the social order as a whole" (Meek 1962, 19).

Quesnay's interests in economic phenomena surfaced late in life, in his sixties, when he recognized that the king's bodily health was linked to economic prosperity. Quesnay never wrote a treatise on economics, only short works in the form of dialogues, maxims, encyclopedia entries, and, most famously, a series of *tableaux économiques* refined over the course of several years, 1758–67 (see Meek 1962).[5] He was intimately involved in the writings of his disciples, notably Mirabeau's *Philosophie rurale* (1763) and Mercier de la Rivière's *L'ordre naturel et essentiel des sociétés politiques* (1767). Indeed, Meek insists that the core set of ideas of Physiocracy were entirely composed by Quesnay and that even Mirabeau's 1763 text was mostly penned by his mentor (Meek 1962, 38).

As several scholars have noted, Quesnay cottoned readily to the analogy between the body politic and the human body. His discernment of the circulation of goods between three sectors—the landowners, the artisans, and the farmers—was directly inspired by his knowledge of human physiology. The fact that the blood system is bifurcated was not lost on him. As Foley has observed, both systems "begin with an initial division of the circulating medium into two separate and equal flows. The blood goes equally to the upper and lower parts of the body, just as currency goes in equal amounts to the productive and sterile classes" (Foley 1973, 134). The flow of blood into the smaller arteries, and its subsequent return in the veinal system, is mirrored by the ongoing division of the original "advance" zigzagging back and forth between the sectors and the eventual return in monetary terms.

Foley suggests that the artisanal sector, insofar as it is unproductive, serves as a "passive conduit for money and goods," much like the veins (Foley 1973, 135). By contrast, the agrarian sector, like the arterial system, definitely enhanced and accelerated the goods involved. There was a "produit net" in this sector, the means to vitalize the economy and keep it growing. Moreover, production took place at the nodes where the two systems meet, just as the capillaries not only join the veins and arteries but also nourish the bodily tissues (Foley 1973, 135–36). The landowners play the most critical role in the picture insofar as they advance the capital necessary for the entire process of production. They are analogous to the heart in that they give the initial impetus, which then courses throughout the entire body. One of the great insights made by Quesnay was the recognition that a single transaction is followed by a series of iterations, much like Keynes's concept of the multiplier. Partly for this reason, as Gianni Vaggi has emphasized, capital tends to take precedence over labor in the Quesnesian economy (Vaggi 1987, 173). It is that advance of capital, like the initial surge of blood as it leaves the left ventricle of the heart, that keeps the system going.

More recently, Loïc Charles (2003) has shown the possible influence of Gaspard Grollier's rolling-ball clock and hydraulic machine on the visual content of Quesnay's *tableaux,* which were intended as pictures or images of the zigzagging circular flow quite independent of the textual analysis. The periodic requirement of a clock winder also transferred to the role of the landlords who set everything in motion. The fluid metaphor might, thus, have had nothing to do with Quesnay's life as a physician. More likely, Charles suggests, both the clock and the circulation of the blood served a heuristic value in his original conception of the economic flow.

Vernard Foley has also argued that the image of Cartesian vortices played a role in Quesnay's circulatory system of wealth and that Quesnay was much more steeped in Cartesian philosophy than most of his contemporaries (Foley 1973, 145). As several scholars have argued, the exact details of Harvey's system were of lesser importance than the commitment to centripetal configurations (see Charles 2004). Foley has also traced Quesnay's profound commitment to iatrophysical thought, the belief that one can understand physiology in mechanistic terms, for example, treating the heart as a pump. More recently, Jessica Riskin (2002, 2003) has questioned the Cartesian component in Physiocracy, emphasizing instead its antimechanist thrust and its assimilation of subtle fluid theories. She places the Physiocrats' appeals to nature within the context of "sentimental empiricism," by which she means "an inseparable fusion of physical sensation and moral sentiment" (Riskin 2003, 43). Whichever source is correct, the evidence is overwhelmingly in support of the claim that Quesnay's training and prolonged interest in natural philosophy influenced the very structure and theoretical core of his economic thinking.

Most significantly, Quesnay saw wealth as part of nature, as a physical substance that only nature could produce and reproduce. He took issue with Cantillon's belief that labor could create value. Man's efforts were deemed sterile. In producing furniture or clothing, we simply transform matter from one state to another. Nature alone can truly augment our wealth, for, with every seed planted in the spring, we reap two in the fall. It is the sun, the rain, and the soil that create genuine wealth. As Du Pont remarked: "Agriculture is the only human labour with which the Sky cooperates without ceasing and which is a perpetual creation. We strictly owe the net product to the soil, to Providence, and to the beneficence of the Creator, to his rain that beats down and changes it to gold" (quoted in Weulersse 1910, 1:275; translated in Banzhaf 2000, 519).[6] Here was the perpetual motion machine so long sought by natural philosophers! Furthermore, the role of the deity in providing this wondrous system of continual wealth entailed that moral laws be intrinsically part of physical nature. As Quesnay declared in his essay "Natural Right": "I am here taking moral law to mean *the rule of all human action in the moral order conforming to the physical order which is self-evidently the most advantageous to the human race.* These laws [natural and moral] taken together constitute what is called *natural law*" (Quesnay 1765/1962, 53). The moral and the physical were, thus, part of one seamless whole.

Quesnay's views sound somewhat odd to our modern ears, but, in many respects, he has confronted a deep question, one to which there is no satisfactory answer. Where does wealth come from, and how is it created? If it is to be thought of in physical terms, wealth can be augmented only by some other agent, by nature, as Quesnay put it. No other investment of human labor or capital truly yields something from nothing. There is always a sense in which what one puts in is simply transformed but not augmented. Even the labor theory of value treated labor as embodying itself in an object. It creates value but not necessarily wealth. Another way in which Quesnay defended his privileging of agriculture was to note that, in manufacturing, all revenues merely compensate the costs of production. Only in agriculture are there gifts from nature that require no such compensation. A popular account of Physiocracy by Nicolas Baudeau captured it thus: "Consider, Madame, the earth covered with its *natural* productions at the moment of harvest, . . . gather together all that mass of Nature's benefits, received by man in the space of *one year:* that, Madame, is the *annual production* or the *total reproduction,* whose idea . . . is the first foundation of the *tableau économique*" (quoted in Spary 2003, 33).

A slogan popularized by the Physiocrats but dating back to Boisguilbert and Gournay is, "Laissez-faire, laissez-passer," which translates best as, "Let nature take its course." The more one could dismantle specific human customs and regulations that thwarted nature's gifts, the better. Free trade of corn (at least domestically) and a single tax levied on the productive sector appeared to the Physiocrats to best accord with the natural order, presumably because agriculture reaped the gifts of nature. At a more general level, the sovereign was to govern according to preordained law. In a purported dialogue between de la Rivière and Catherine the Great, de la Rivière defined the science of government as "the study of the laws which God has so evidently engraven in human society when He created man. To seek to go beyond this would be a great calamity and a destructive undertaking" (quoted in Spiegel 1991, 198).

Paul Christensen has documented in some detail Quesnay's command of natural philosophy, including his interest in nonmechanical thinking, such as the doctrines of the ether and of subtle fluids. Christensen maintains that Quesnay qua physician had turned against the Cartesians and revived some of the Aristotelian and Galenic ideas, for example, the three souls and the four humors. More specifically, Quesnay first cultivated the idea of physical productivity in his *Essai phisique sur l'oeconomie animale.* As Christensen has argued, Quesnay emphasized the Aristotelian idea of the nutritive soul, which he saw as the central motive force, far more than the rational soul, in both the human body and the body politic. This emphasis in turn highlighted the material nature of the "economy," which in turn bore a strong resemblance to physical nature and the animal oeconomy. Christensen has shown too that the concept of the ether figured centrally in Quesnay's work and that this helps account for the source of productivity in the agricultural sector. Quesnay's ether, a subtle and active stream of particles like fire, kept the system in circulation, both in the sense of material transformations among artisans and farmers and in the sense of bodily functions such as digestion or muscular activity.

Benjamin Franklin's experiments on electricity engendered considerable enthusiasm among the French, including the Physiocrats (see Riskin 1998, 2002). Buffon oversaw the translation in 1752 of Franklin's letters to Peter Collinson. Another lesser-known Physiocrat, Jacques Barbeu-Dubourg, conducted electrical experiments and also translated Franklin's papers, publishing the ones on electricity and political economy together in one volume in 1773 (see Riskin 2002, 114). Franklin's work was admired for its Newtonian spirit, for the manner in which his theoretical claims were bolstered by simple but elegant experiments. Despite opposition from the Cartesian electrician Nollet, Franklin had a greater hold on the French savants. In addition to Buffon, he was endorsed by Du Pont, Daubenton, Needham, and Turgot. Du Pont, for example, recognized in a 1764 address that wealth spreads like electricity, both because it acts like a fluid and because it is common to all who come in contact with it (Riskin 1998, 336). Franklin's ideas were explicitly promoted by the Physiocrat's journal, *Éphémérides*. When Franklin visited France in 1767, he found many devoted followers, *les Franklinistes* as they were known. By 1769, Franklin himself was converted to Physiocracy, accepting the preeminence of agriculture and the view that only God, via nature, creates true wealth (Riskin 1998, 333). When Franklin died in 1790, the National Assembly declared three days of national mourning; Condorcet gave the *éloge*.

As Riskin (2002, 120) argues, both Franklin and the Physiocrats placed great weight on the properties of conservation and the maintenance of a balance in nature.[7] For Franklin, the Leyden jar transferred the electric fluid from the outer surface to the inner one, to maintain a balance. One could also think of it as being charged plus or minus and of a transfer of the fluid to another body as rendering the host body neutral. Franklin's leitmotiv was that of a bookkeeper, whether in his economics or in his concept of plus and minus electricity (see Riskin 1998, 316). Even more profoundly, Franklin's theory of negative and positive electricity was "shaped by a mode of moral reasoning in terms of conservation of harms and goods" (Riskin 2003, 57). Franklin's analogical trade between physics, economics, and morals matched the predilections among the Physiocrats to draw on concepts and imagery in physiology and mechanics. The latter group, however, fostered the more significant belief that nature alone can create economic wealth.

OTHER EIGHTEENTH-CENTURY CONTRIBUTORS

Other French savants who were intrigued with Physiocracy but were not official members of the group were Étienne Bonnot de Condillac (1714–80), Antoine Lavoisier (1743–94), and Anne Robert Jacques Turgot (1727–81). Condillac and Lavoisier are better known for their contributions to natural philosophy and were generally seen at the time as helpful lieutenants to Turgot in his efforts at economic reform. Condillac spearheaded the movement to reform scientific language, and Lavoisier followed suit in his efforts to revise chemical nomenclature. Both saw the importance of injecting rigor and precision into economic language as well. On the opening page of his one economic text, *Commerce and Government* (1776), Condillac declared: "Each science requires a special language, because each science has ideas which are unique to it. . . . That is the position of Economic Science, the subject of this very work" (Condillac 1776/1997, 78). Condillac was far less inclined than his contemporaries to ground the economic order in nature, although he still retained an emphasis on agriculture as the true source of wealth. As a result, he also endorsed the Physiocrats' policies of a single tax and free trade for grain. But he also developed a theory of value that highlighted the role of utility, and, for this, he is often seen as more aligned with Galiani and Turgot (see Eltis 1995). Moreover, Condillac believed that convention and artifice are woven right through society and, thus, that social reform was more about addressing institutional repair than full-fledged dismantlement.

In his economic inquiries Lavoisier focused mostly on fiscal reform, to which he brought the passion for quantitative measurement that was so significant in his chemical investigations (Poirier 1993, 43). The conservation of matter was also a paramount principle in his work with gases and had a direct counterpart in his own construal of the circulation of goods between producers and consumers (Poirier 1993, 267). He also discerned that respiration is a form of combustion and, thus, an exchange of gases, of oxygen for carbon dioxide. He directly imported notions of consumption, production, and exchange into his study of the "animal economy," as Norton Wise has observed (Wise 1993, 222). He also wove the concept of a balance into both his chemical and his economic inquiries. He drafted balance sheets for the production and circulation of agricultural products in a fashion parallel to his analysis of caloric, the fluid of heat. Moreover, in his *éloge* for Colbert (1771), he drew a direct analogy between money and the caloric fluid: "Hard currency is a fluid, not, it is true, so mobile as water, but which with time necessarily takes its level" (quoted in Wise 1993, 252, from Lavoisier 1771, 116).

Simon Schaffer has drawn attention to the fact that both Lavoisier and another celebrated physicist, Charles Coulomb, investigated the measurement of labor in mechanical terms (see Schaffer 1999, 134). Lavoisier's analysis of bodily functions like respiration closely mimicked his quantitative studies of human labor more generally. The tradition of treating the human body as a machine had been well established since Julien Offray de La Mettrie's *L'homme-machine* (1748). Both Adam Smith and Quesnay were taken with La Mettrie's efforts to mechanize the animal oeconomy and extend this into the human realm (Schaffer 1999, 142–43). There are, thus, mechanical as well as organic antecedents to Lavoisier's studies.

Turgot was affiliated with the Physiocrats, although he insisted that he did not officially belong to the group. The reasons for this were partly temperamental (he chided his good friend Du Pont for sitting at the feet of his master, Quesnay [see Riskin 2003, 62]) and partly strategic (he held various government offices and was, thus, concerned with public appearances). But there were also some fundamental differences; Turgot objected strongly to the physiocratic positions on despotism and fiscal policy (see Poirier 1993). Turgot's economic analysis is also deviant, notwithstanding the strong endorsement of laissez-faire notions and the emphasis on balance in his one book, *Réflexions sur la formation des richesses* (1766). On becoming controller-general in 1774, he enlisted the close counsel of Du Pont and took under his employ Quesnay's own grandson. We know that Turgot and Du Pont remained on very close terms for several decades, even to the point of assisting one another with financial and political challenges in the wake of the Revolution. Another close associate of Turgot's was the Marquis de Condorcet (1743–94), who wrote more generally about *la science sociale,* a term that he first coined, and is revered for some interesting breakthroughs on voting patterns and applications of probability theory. A nascent positivist, Condorcet shared with Turgot a strong conviction in the natural progressive order of society that could be achieved through rational means such as "social mathematics" and the reform of language (see Baker 1975; and Gillispie 1980). Turgot's own efforts to reform the French economy were short-lived. He lost the post of controller-general by 1776, a loss that coincided with the demise of Physiocracy. Condorcet met an even nastier end, forced to go into hiding in 1793 and dead within days of his arrest the following year.

Turgot had studied theology and law at the Sorbonne, but he also acquired a solid grounding in natural philosophy. In the history of the physical sciences, Turgot made a significant breakthrough with his concept of *expansibilité*. He recognized that, in principle, all substances, not just air, would enter into a vaporous state if heated long enough and that expansion could go on indefinitely. The concept of *expansibilité* owed much to Hales's idea of fixed air and to Black's idea of latent heat. For Turgot, as heat is added, it wedges itself between the particles of a substance, thus causing the expansion. Lavoisier made great use of Turgot's idea insofar as it helped him grasp that the different physical states—solid, liquid, and gaseous—are independent of the chemical composition of a given substance (see Hankins 1985, 87–88, 97–100). Turgot also embraced the idea that human history evolved by stages, the classic ones of hunting and gathering, animal husbandry, cultivation, and, finally, commerce (see Meek 1971). There is no evidence that the two types of stadial thinking were linked in Turgot's own conceptual genesis. The so-called four stages theory of human history was popular at the time, especially among Scottish moral philosophers such as John Millar and Adam Ferguson. The only point that one can make reliably is that Turgot cultivated interests in natural philosophy, and, thus, there is prima facie reason to believe that these ideas infiltrated into his central efforts at economic theory and policy.[8]

During the 1750s, Turgot worked for Gournay, who at that point served as the *intendant du commerce* and was one of the major intellectual forces in French economic thought in the first half of the century. Turgot's economic writings appeared contemporaneously with those of Quesnay. He wrote several articles for the famed *Encyclopédie* (1756–57), including two on economic subjects (one on fairs and one on markets), the *Éloge de Gournay* (1759), and other miscellaneous notes and papers. Turgot formed a friendship with David Hume while he was visiting Paris in the 1760s and also had the opportunity to meet Adam Smith. His *Réflexions* was the result partly of Hume and Smith's inspiration and partly of his acquaintance with the by then vigorous group of Physiocrats.

The *Réflexions* is characteristically deep and succinct. In it, Turgot, as did Quesnay, divides society into economic classes, initially just two, the cultivators and the artisans. He then tells a quasi-Lockean tale about the emergence of money as a "pledge" and the eventual formation of a landowning class. Even though all goods have two of the essential properties of money, those of "measuring and representing all value," only gold and silver have the requisite natural properties to serve as universal money (Turgot 1770/1898, 36, 39). The formation of a class of proprietors is, in Turgot's view, inevitable and, more important, self-sustaining through the mechanism of inheritance. Much of what transpires for Turgot does so by "the nature of things." He is, thus, a strong adherent, like Du Pont, to the view that economic properties are part of the natural order.

Although Turgot grants industry the right to productivity, agriculture is still privileged: "It is the earth which is always the first and only source of all wealth" (Turgot 1770/1898, 46). Even other forms of wealth or capital are always equivalent to some piece of land (50). All human production requires an advance, which exists only because of the fecundity of the soil. The only genuine revenue is the net produce of the land, and the land alone has "furnished all the capitals which make up the sum of all the advances of agriculture and commerce" (96–97). The reason stems from the fact that only nature makes a "pure gift" incommensurate to anything that man contributes: "Nature does not bargain with him [man] to oblige him to content himself with what is absolutely necessary. What she grants is proportioned neither to his wants, nor to a contractual valuation of the price of his days of labour. It is the physical result of the fertility of the soil, and of the wisdom, far more than of the laboriousness, of the means which he has employed to render it fertile" (9).

Two points bear noting here. First, there is no contract with nature, no just exchange. Nature gives but does not take back. Second, wealth resides in the "physical result of the fertility of the soil," not in labor. Turgot thus locates wealth, as had his immediate predecessors, in physical nature. It appears to grow in human hands but is, in fact, a very separate process of "generation," much like the growth of plants (see Fontaine 1992, 78).

As Philippe Fontaine has observed, Turgot borrowed Hume's distinction between physical and moral causes, whereby the latter were essentially identical to the realm of human agency. But Turgot put his own twist on the matter. While Turgot would have little truck with Montesquieu's extreme physicalism, he nonetheless allowed that even the inner recesses of the mind are subject to physical causes. However, because we observe the moral phenomena of motives and actions, Turgot rejected a thoroughgoing physical determinism (Fontaine 1997, 4). Nevertheless, these moral causes are resonant with the physical causes and act in the same law-like fashion. In that sense, Turgot, as we will see, reads like Hume. As Fontaine notes, the moral causes are located more in institutions and the broader social fabric than in the individual mind. Institutions such as money come to pass independently of human deliberation. The human will is, thus, subordinate to external forces.

NINETEENTH-CENTURY DEVELOPMENTS

Political economy in France after the Revolution was much more sectarian than its counterpart in England. Condorcet and Destutt de Tracy (1754–1836) forged with Jean-Baptiste Say (1767–1832) a school of *idéologues* that sought to advance considerably the cause of liberal trade and government (see Forget 1999). Positivism was explicitly launched by Auguste Comte (1798–1857), with his six-volume *Cours de philosophie positive* (1830–42). With the exception of Adam Smith's work, Comte held political economy in contempt and did not grant it the status of a separate science. By the first half of the nineteenth century, economic inquiry had clustered into three distinct strands. The first strand is the classical political economy exemplified by Say and Frédéric Bastiat (1801–50), who were known politically for championing liberal trade policies. The second is the socialists, led most prominently by Henri de Saint-Simon (1760–1825), Charles Fourier (1772–1837), Simonde de Sismondi (1773–1842), and Pierre-Joseph Proudhon (1809–65) (see Ege 2000). The third—which, taking inspiration from Condorcet, sought to mathematize economics—included Nicolas-François Canard (1750–1833), A. A. Cournot (1801–77), and a group of engineers, most notably Jules Dupuit (1804–66) of the École des Ponts et Chaussées (see Ménard 1978; T. Porter 1995, chap. 3; Mosca 1998; Vatin 1998; and Ekelund and Hébert 1999).

Say is generally regarded as the most influential French economist of the first half of the nineteenth century and the one who most clearly aligned himself with the classical school in Britain. Recent scholarship has done much to show that Say was not simply a disciple of Adam Smith's, as he had earlier been considered to be (see Forget 1999, 107). His *Traité d'économie politique* (1803) was, on its publication in France, promptly translated—selling widely in Britain and North America—and sparked debate over what came to be known as Say's law, a set of claims contending that a general glut of goods is impossible. The law is essentially a principle of conservation, that there is no loss of purchasing power in the economy. A more significant point to extract from the *Traité* is that in Say, as in Ricardo (as we shall see), is found a full-blown concept of an economy. Say was explicit about aggregate demand and supply and about aggregate savings and investment, to use modern terminology. His analysis also highlighted the concept of an equilibrium in economic discourse.

For Say it was a given that the science of economics must be grounded in the nature of things, especially the laws of human physiology. As Evelyn Forget has argued in some detail: "*Idéologie* drew its intellectual credibility from the advances in the life sciences, particularly physiology" (Forget 1999, 66). Together with Destutt de Tracy, Say forged a theory that took direct inspiration from the physiological writings of François-Xavier Bichat and P.-J.-G. Cabanis. Elements of their studies of the nervous system and overall functional harmony in the body were projected onto the image of the body politic. Much was made of the developmental properties of the social organism. Societal health was to be improved via moral education in deference to the fundamental operation of sympathy. The most prominent public health reformer of his time, Louis René Villermé (1782–1863), took considerable inspiration from Say and his fellow *idéologues,* both for their commitment to laissez-faire apologias and for their belief in the inherent productive capacity of industry (see Coleman 1982). Villermé firmly believed that only idleness perpetuated poverty and, thus, like Say and Destutt de Tracy, emphasized moral improvement.

With hindsight, all these strands can be seen to have fed directly into the work of the most eminent of nineteenth-century French economists, Léon Walras (1834–1910). Walras absorbed socialist tenets from his father, Auguste Walras (1801–66), and the Saint-Simonians as well as the mathematical-engineering tradition via his education at the École des Mines. Walras's admiration for Cournot in particular was part and parcel of his lifelong mission to mathematize economic theory. By the 1870s, economics had become a respectable science. Chairs in the subject had been established at every university in France, mostly because of Say's influence. Much to his dismay, Walras was able to secure a post only in Switzerland, at the Acadamié de Lausanne (later the Université de Lausanne), and was unsuccessful in his efforts to return to France.

Walras is not an easy figure to distill. He was much more steeped in ruminations about human needs and desires than has generally been recognized by Walras scholars, so bent are they on rendering sensible his theory of general equilibrium.[9] In several early papers, Walras reflected on the nature and methodology of science, both natural and moral, and put most emphasis on the phenomena under investigation themselves dictating the boundaries. The most fundamental distinction is between human and natural facts, the latter pertaining to plants, animals, and minerals. Walras presents a visual chart of his categories and is explicit that humans and nature do not intersect. There is pure natural science for him, although the study of humans can branch off into both the moral sciences and applied natural science with practical ends (see Jolink 1996, 78; and Lallement 2000, 450–52). Walras, in short, gives us the ontological divide addressed in chapter 1, whereby the economists (the moral scientists) and the physicists (the natural scientists) have divided the world into nonintersecting domains. It is not certain when he wrote these papers, but they most likely date from the 1860s. This would put Walras in the wake of Mill's fundamental divides and, thus, bolster my overall thesis that the denaturalization of the economic order had come to pass by the mid-nineteenth century.

Walras's *Éléments d'économie politique pure* (1874) was one of the three critical texts that unleashed the so-called Marginal Revolution, from which issued neoclassical economics. Economists still refer to Walrasian economics as the reigning paradigm. By the time Walras's views made their mark, roughly at the end of World War I, the transformation that I have characterized as the denaturalization of the economic order had run its course, propelled by the secular efforts of Condorcet, Comte, and Walras.

Yves Breton (1998, 404) has argued that marginalism was slow to gain recognition in France and that, even for the first twenty or so years of the twentieth century, it remained controversial. For much of the nineteenth century, French economics lay in the shadow of the British, dominated by the work of Ricardo, Malthus, Mill, Jevons, Edgeworth, and Marshall. Furthermore, as Joseph Schumpeter observed, the strength and quality of the second tier of nineteenth-century economists greatly contributed to the "unrivaled prestige that English economists then enjoyed" (Schumpeter 1954, 382–83, 757). A recent three-volume collection, *Nouvelle histoire de la pensée économique* (Béraud and Faccarello 2000) gives similar weight to the British economists in its overview of pre-neoclassical theory. A chapter on the French socialists and the entry on Say in the chapter on the second generation of classical economists (Ricardo and Malthus) are the sole pages to counteract the view that the "Age of Capital" clearly belonged to the British (see Hobsbawm 1975, 316). Only with the widespread acceptance of Walrasian general equilibrium theory ca. World War I did France regain the international influence it once had with Quesnay.

CHAPTER FOUR

Hume's Political Economy

'Tis evident, that all the sciences have a relation, greater or less, to human nature; and that however wide any of them may seem to run from it, they still return back by one passage or another. Even Mathematics, Natural Philosophy, and Natural Religion, *are in some measure dependent on the science of* MAN.
—David Hume, *A Treatise of Human Nature*

Among Enlightenment philosophers, David Hume (1711–76) provided one of the most extensive arguments for the cultivation of "the science of MAN" (Hume 1739–40/2000, 4). Even mathematics and natural philosophy depend on our human faculties, as is made plain in the epigraph to this chapter, taken from the introduction to the *Treatise of Human Nature* (1739–40). As part of his efforts to develop the moral sciences, Hume cast aspersions on those engaged in the natural sciences, attacking everything from their inability to know the inner constituents of matter and force to their propensity for becoming overly enamored of the efficacy of a single, purportedly universal principle such as gravitational attraction (Hume 1739–40/2000, 46, 107; Hume 1777/1985, 159–61). Much of book 1 of the *Treatise* leaves one wondering whether knowledge of the natural world is possible at all.[1] Like Vico and Montesquieu before him, Hume took a critical stance toward the natural sciences that paved the way for the cultivation of the moral sciences, as laid out in books 2 and 3 of the *Treatise,* his essays on political economy, and his monumental *History of England* (1754–62).

Although Hume's primary focus was directed toward human nature, I will argue here that the broader framework of his inquiry was the natural realm and that his knowledge of natural philosophy seeped into his moral philosophy and, more specifically, his political economy.[2] My argument is reinforced by noting that, for Hume, a typical event includes causal links within both the natural and the moral realms and there are few isolated moral phenomena. An understanding of economic phenomena such as money and prices would, thus, necessitate some attention to the physical setting offered by nature. As Hume observes, the world had been designed such that its diverse "geniuses, climates and soils" ensure "mutual intercourse and commerce" (Hume 1777/ 1985, 324, 329). A richer understanding of physical nature would, thus, facilitate a deeper understanding of the moral realm, just as natural philosophy feeds on a fuller understanding of human nature.

Drawing on the work of Michael Barfoot and others, I will show that Hume was well acquainted with natural philosophy and that there are aspects of his economic thought that reflect that knowledge. Barfoot has conjectured that Hume's study of hydrostatics played a role in his emphasis on the flow and circulation of money, particularly in "Of the Balance of Trade" (see Barfoot 1990, 165). In that essay, Hume drew explicit analogies between money and water, making much of the tendency of water to reach its own level. All this points to Hume's appreciation for the natural setting of economic phenomena. But I wish to go even further. In the essay "Of Money," there is no explicit mention of water. Rather, money is treated as a fluid in more general terms. Hume may well have developed his insights by drawing on recent experiments on the so-called electric fluid. I have not been able to locate statements whereby Hume explicitly links electricity and money, and my argument is, thereby, the weaker, drawing only on circumstantial evidence and, thus, reaching only a tentative conclusion. Nonetheless, I believe that it would be difficult to explain some of the more striking aspects of Hume's monetary theory without making these connections.

Andrew Skinner has observed that the unifying theme of Hume's economic writings is the appeal to history, or what he calls Hume's "historical dynamics" (Skinner 1993b, 244, 248). I will take this one step further and argue that Hume's historical approach to economic phenomena parallels the nascent evolutionary thought of his day. Hume's identification of economic cycles in terms of centuries is explicitly linked to growth and decay in physical nature, a temporal reach that was also manifest in the natural history of Buffon or Hutton. Hume also elevates the reason of animals and downgrades that of humankind, suggesting that, in that respect too, we are closely aligned with the natural world. This also serves to reinforce his overarching secularism. These themes will be explored below.

Alas, the scholarly tradition has generally separated Hume the philosopher from Hume the economist. Duncan Forbes's magisterial analysis of Hume's political philosophy deliberately set aside the economic writings because of a limited command of the subject (see Forbes 1975). Likewise, those who focus on the history of economics tend to dodge Hume's philosophical contributions, in part because they are so formidable. Ironically, despite the lengthy introduction by Eugene Rotwein (1970) positioning Hume in the broader philosophical milieu, most historians of economics have taken the collection of nine essays as the exclusive source on the subject. Eight of these essays were first published in 1752 as *Political Discourses;* the ninth, "Of the Jealousy of Trade," was published in 1758. But, in each case, these essays were published with others, mostly on political subjects, for example, "Idea of a Perfect Commonwealth" and "Of the Coalition of Parties." In short, there is no reason to suppose that Hume saw himself as issuing a set of essays on *economics,* a term that did not yet have common currency, at least in English. Rather, there are good grounds for viewing Hume's essays, which in 1758 he labeled *moral, political, and literary,* as a unified attempt to give specificity to his earlier proposal, in the *Treatise,* for cultivating the science of man.[3]

Moreover, the emphasis of late among Hume scholars has been that of coherence, both within the *Treatise* itself and between it and the political essays (see Baier 1991; and Haakonssen 1994).[4] I am, therefore, inclined to emphasize coherence and consistency among Hume's works while recognizing that no profound writer can ever be fully consistent. In support of my predilection is the fact that Hume was remarkably vigilant, revising and polishing his works, especially his essays, from the time they were issued up until his death (see Hume 1777/1985, xiv). From 1758 through 1777, the essays appeared in eight editions, Hume's changes ranging from different wordings, to different groupings of the essays, to the addition/deletion of entire essays.

THE NATURAL AND THE MORAL

In the *Treatise,* Hume defines nature as those operations that are "independent of our thought and reasoning" (Hume 1739–40/2000, 113). Our knowledge of physical nature is, therefore, always circumscribed by the chasm that separates our minds from it. Moreover, the physical forces that keep the swirl of matter in motion are by their very nature unobservable and, thus, beyond our reach. We might well observe the continuous succession of objects, but "if we go any farther, and ascribe a power or necessary connexion to these objects; this is what we can never observe in them, but must draw the idea of it from what we feel internally in contemplating them" (114). As a result, our reasoning about nature is on a par with our reasoning about the moral or human realm: "The same course of reasoning will make us conclude, that there is but one kind of *necessity*, as there is but one kind of cause, and that the common distinction betwixt *moral* and *physical* necessity is without any foundation in nature" (115). Much of Hume's metaphysical analysis is intended to level our epistemological command of the physical world and the human world.

In his *Enquiry concerning Human Understanding* (1748), Hume maintains that, in many of our inferences, we move unwittingly from natural to moral phenomena, that those inferences form "one chain of argument" and must, therefore, be "of the same nature, and derived from the same principles" (Hume 1748/2000, 69). He gives the graphic example of a prisoner facing execution. That man will draw the same sound inferences regarding the inevitable actions of his jailor, the soldiers attending his march to the scaffold, and the executioner as he will regarding the physical laws that govern his beheading. More important, the "mind feels no difference" in moving from one link in the chain to the next, despite the fact that some are moral and some natural (Hume 1748/2000, 69). If economic phenomena are mostly drawn from the moral realm, they are, thus, readily interspersed with events in physical nature.

Hume's central point, by now a famous one among scholars, is that we are warranted in inferring the constancy of human behavior. Human beings may not act as consistently as billiard balls, but, for the most part, people follow predictable paths. Hume suggests that the irregularities that we observe and come to expect in human action are comparable to those in the weather. Both human action and the weather exhibit fluctuations, but the patterns are unmistakable. There is undoubtedly an underlying set of "steady principles" that govern those phenomena, the "winds, rain, clouds," however difficult they are to discern (Hume 1748/2000, 67). As Jan Golinski has remarked of mid-eighteenth-century science: "The weather came to be viewed less as a vehicle of providential intervention through singular events and more as a natural process accessible to human knowledge" (Golinski 1999, 86). Golinski charts the widespread manufacture and obsessive use of the barometer in the 1730s and 1740s as a central means of rationalizing the weather. As part of this cultural movement, Hume compared the reliability of reading the interest rate to that of reading a barometer: "If we consider the whole connexion of causes and effects, interest is the barometer of the state, and its lowness is a sign almost infallible of the flourishing condition of a people" (Hume 1777/1985, 303). While we should not read too much into this analogy, it does suggest that

Hume regarded economic phenomena as partaking in the relative stability of physical nature.

Furthermore, Hume puts considerable weight on the "constant character of human nature" (Hume 1748/2000, 67) while of course recognizing that human nature has some capacity to evolve and undergo refinement over time, depending on the customs and economic well-being of the age. Hence his often-cited invitation to compare the "sentiments, inclinations, and course of life" of his contemporaries to those of the people of ancient Greece and Rome (Hume 1748/2000, 64). It is precisely this constancy that allows us to pursue the sciences of history, politics, and morals. For Hume, this is indisputable. The "conjunction between motives and voluntary actions is as regular and uniform, as that between the cause and effect in any part of nature" (67). History appears to repeat itself for Hume: "Mankind are, in all ages, caught by the same baits: The same tricks, played over and over again, still trepan them" (Hume 1777/1985, 363). Contrary to Descartes, there is no more free will in the mind than in the body: "The fabric and constitution of our mind no more depends on our choice, than that of our body" (168). While Hume argues persuasively that causal ascriptions cannot be demonstrated to be true, insofar as we do engage in this practice we are equally justified in the natural and the moral realms. His analysis of causation, of freedom and necessity, implies that human actions are as law governed as anything in nature.[5]

Most often, Hume uses the word *natural* to mean "that which is normal or commonplace." But he also draws an important distinction between the natural and the artificial in the case of duties and virtues. At first reading, he seems to mean the following, at least for duties. Those duties that stem immediately from instincts—the duty to love one's children, for example—are deemed natural, while those that must be reasoned through—that to act justly, for example—are deemed artificial. Much more ink has been expended on what Hume meant by this distinction when applied to the set of virtues (see Baier 1991, chap. 9; and Gauthier 1992). It may be understood that a purely natural virtue is one the realization of which requires no additional motivation or act of the will. Yet, under closer scrutiny, the distinction is not hard and fast but, rather, one of degree. To settle this issue here is not particularly pressing. Suffice it to say that the fact that Hume drew the distinction meant that he was aware of the importance of positioning human agency vis-à-vis the natural realm.

Hume emphasized the role of instincts and passions in governing human action. Instincts are, of course, reminiscent of animal activities, while the passions are subject to the "dictates of nature" (Hume 1777/1985, 141). For example, nature has implanted sexual passions in us, and avarice is ever present. Many passions, such as ambition and gratitude, are "of a very stubborn and intractable nature" (97; see also 113, 131). Reason is no savior, as we are reminded by one of Hume's most infamous remarks: "Reason is, and ought only to be the slave of the passions" (Hume 1739–40/2000, 266). Moreover, the passions—for example, the love of another person or an appreciation for beauty—are reducible to the "structure and constitution" of the mind (Hume 1777/1985, 163). They are, thus, part of our physiology or, more precisely, of the three laws of association that channel our every thought, as adumbrated in the opening sections of Hume's *Treatise*. Ultimately, there is no escape from the forces of the passions.

Although there is, thus, a strong sense in which we humans are seen by Hume to be seamlessly joined with things natural, there is another sense in which nature stands apart and is superior to us and eludes our grasp (Hume 1777/1985, 138). Human art and artifice have but a poor grasp of nature's perfections and inner workings: "Art copies only the outside of nature, leaving the inward and more admirable springs and principles; as exceeding her imitation; as beyond her comprehension" (158). In this sense, Hume creates a picture whereby humans are beholden to nature. Our admiration and reverence spring forth unreservedly.

Nature is also providential. She gives us clues by which we might fathom her ways and, thereby, find the path to virtue. Nature has a plan, to "make us sensible of her authority" (Hume 1777/1985, 163). If we do not undertake sufficient industry and application, then, as befits the station of our species, she "revenges herself in proportion to our negligent ingratitude" (148). Hume entreats us to use our intelligence and, thus, rise above the lowest level of animality (147). Nature's prodigality also allows room for error; nature "has provided virtue with the richest dowry" such that one may "pay to virtue what he owes to nature" (153). Nature is, thus, benevolent as well.

Notwithstanding his avowed anticlericalism and purported atheism, Hume made numerous appeals to Providence and the "Author of Nature" in his essays and treatises. In keeping with the views of his time, he sustained a deeply rooted belief in a natural order. He even suggests in places that such order was the work of the deity: "There surely is a being who presides over the universe; and who, with infinite wisdom and power, has reduced the jarring elements into just order and proportion" (Hume 1777/1985, 154). Although it is safe to assume that Hume had broken with the scriptural account of creation, it is still a matter of debate whether he embraced a fully materialistic account of the world's origins.[6]

Whatever the case, Hume was one of the most secular thinkers of his time. For example, he saw human beings as more closely connected to physical nature than to anything spiritual. Man was much more like an ape than an angel: "Man falls much more short of perfect wisdom . . . than animals do of man" (Hume 1777/1985, 83). Indeed, Hume was one of the first to suggest that humans and animals think alike, that we both form our beliefs through repeated trials or a process of experimental reasoning (see Hume 1748/2000, sec. 9).[7] Even our basic belief in causation is formed by custom and habit, which stem, not from reason, but from instinct (Hume 1748/2000, 44–45).

Man also resembles animals in terms of the balance struck between passions, reason, industry, and play. Hume grants animals a wide range of passions—love, hatred, sympathy, grief, and even malice—implying that we share much with them. His willingness to put man on the same footing as animals owes much to his secular orientation. In his posthumous essay "Of Suicide," Hume suggests that man and animals are cut from the same cloth and, in that sense, are of equal (in)significance: "The lives of men depend upon the same laws as the lives of all other animals; and these are subjected to the general laws of matter and motion" (Hume 1777/1985, 582). Providence has arranged things such that each animal is entrusted with its own care. A human being may, thus, end his life as freely as he chose to preserve it. Neither act will disturb the course of nature: "The life of man is of no greater importance to the universe than that of an oyster" (583).

Hume nonetheless contrasts the moral realm to the physical world and claims that physical causes are not always the predominant factor in determining a given "moral phenomenon." For example, he contested Montesquieu's theory that climate determines national character, citing the Chinese as a counterexample.[8] But, whether the phenomenon be moral or physical, it operates in accordance with the same patterns of causation. Both types of phenomena, he argues at length, exhibit patterns that enable us to form habits and beliefs governing the course of every action. For Hume, there are stable and robust character traits that stem from external conditions: "A soldier and a priest are different characters, in all nations, and all ages; and this difference is founded on circumstances, whose operation is eternal and unalterable" (Hume 1777/ 1985, 199). It is this stability that motivated his conviction that there are laws in the moral sciences and that those laws are as constant as those in the physical sciences. For example, we no more expect our dinner guest to stab us and steal our silverware than we do our house to topple in an earthquake (Hume 1748/ 2000, 69).

For Hume, as has already been noted, there is no more free will in the mind than in the body. He allows meager room for character reform; most, if not all, of our character is determined by factors beyond our control. As Paul Russell has persuasively argued, even when character change is achieved, it is mostly, if not entirely, due to factors that lie beyond the individual sphere of choice (Russell 1995, 130). In this respect, Hume has leveled the natural and moral realms. Both are governed by a finite set of laws and are, thus, part of one unified order.

HUME'S KNOWLEDGE OF NATURAL PHILOSOPHY

Let me now turn to the question of Hume's command of natural philosophy and the extent to which this shaped his writings on political economy. The case that he was well versed in and influenced by natural philosophy is a harder one to make with Hume than with Adam Smith (as we will see in the next chapter). But there is little question that Hume's education and adult intellectual life matched Smith's when it came to contact with the natural sciences. Hume had taken Robert Steuart's course on natural philosophy while a student at the University of Edinburgh in 1724–25. We know that this course emphasized the experimental sciences, particularly pneumatics and hydrostatics.[9] Insofar as mathematics was a prerequisite for Steuart's course, there is good reason to suppose that Hume studied that subject at an advanced level. Hume was later to associate closely with Colin Maclaurin, one of the most renowned mathematicians of the time, suggesting at least no aversion to the subject. Moreover, in the *Treatise,* Hume wrote confidently about the subject of geometry, for example, the troublesome parallel postulate; the fact that among his manuscripts were found two unpublished essays on mathematics, one by Robert Wallace, one possibly by Hume himself, also suggests that he understood Newtonian fluxions.[10]

Roger Emerson and Paul Wood have both argued for the strong scientific bent of the Scottish Enlightenment philosophers, including Hume. The short-lived Select Edinburgh Society (1754–64), of which Hume, Smith, and Cullen were founding members, is cited as a case in point. The society's mandate was distinctly pragmatic, with applications of scientific methods to such fields as agriculture, engineering, and medicine. Cullen, who helped build the University of Edinburgh's medical school into the most prestigious in Europe, was also personal physician to both Hume and Smith. It is likely that the friendship with Cullen, and later those with Joseph Black and James Hutton, provided Hume with considerable information about contemporary developments in natural philosophy.

Hume's election to the post of (joint) secretary of the Philosophical Society of Edinburgh is another indication of an interest in natural science. The society had originally been created in 1737 to bring together physicians and naturalists but, with time, also welcomed men of letters at large. Experimental demonstrations were occasionally conducted at society meetings. Up until the early 1750s, the society oversaw publications exclusively on medical investigations, but then, under Maclaurin's prompting and Hume's supervision, expanded its purview to include essays on natural philosophy. This transpired during Hume's tenure as secretary (1751–63). He coedited (with Alexander Monro) two volumes of the *Essays and Observations, Physical and Literary* in 1754 and 1756. A third volume, published in 1771, contains a number of pieces from the early 1760s that Hume had most likely edited (see Wood 2002; see also Mossner 1980, 257–59). The unsigned preface, almost certainly by Hume, notes that the volume explicitly excludes "the sciences of theology, morals, and politics" on the grounds that they "will for ever propagate disputes" (Philosophical Society of Edinburgh 1754–71, 1:v–vi). Given the society's strongly scientific orientation, it would have been odd for Hume to spend his time there, let alone serve on the executive, if his interests were exclusively in the moral sciences.

Barfoot has argued that Hume was well versed in the scientific culture of his age. Some prominent examples of this are his analysis of the vacuum and of the divisibility of matter in book 1 of the *Treatise*. Hume also praised Boyle's insights on hydrostatics (Barfoot 1990, 163) and acknowledged the unparalleled genius of Newton and Galileo (Hume 1777/1985, 550). As James E. Force has observed, Hume refers to Newton in every one of his books and in some of his essays as well (see Force 1987, 169–78). While most of the references do not evince great detail, there are one or two that do. In the *Enquiry concerning Human Understanding*, for example, Hume makes explicit reference to Newton's theory of gravitation, to his "*vis inertiae,*" and to "an etherial active fluid" (Hume 1748/2000, 58n), suggesting that he was well acquainted with Query 31 of Newton's *Opticks*. In volume 6 of the *History of England* (of which the last two volumes were completed first), Hume draws attention to the limitations of the new science: "While Newton seemed to draw off the veil from some of the mysteries of nature, he showed at the same time the imperfections of the mechanical philosophy; and thereby restored her ultimate secrets to that obscurity, in which they ever did and ever will remain" (Hume 1778/1983, 6:542). It was commonplace for European savants of the time to pay tribute to Newton, so, in that respect, these references could mean little. The fact that Hume attends to the incompleteness of the Newtonian philosophy, however, suggests a deeper grounding in the subject.[11] Donald Livingston has argued that Hume rejected Newton's theory of gravitation because it retained its metaphysical status (see Livingston 1984, 170–71). A close reading of the *Treatise* suggests that Hume was more aligned with Berkeley's skeptical stance toward Newtonian physics and corpuscularianism than most of his contemporaries (Hume 1739–40/2000, I16).

In support of Hume's scientific credentials, there are his many thought experiments to appreciate. In his *Treatise,* for example, the discourse on the concept of a vacuum strongly suggests a close acquaintance with previous thought experiments on the same subject (Hume 1739–40/2000, 40–43; Barfoot 1990, 172–81). This is a mode of reasoning that flourished in the early modern period in the hands of Galileo and Descartes. Many thought experiments offer a clear demonstration by exploring an implausible, if not impossible, series of events. Most, if not all, cases of economic analysis involve abstraction, the distortion of reality either by idealization or by oversimplification, yet these manipulations in themselves do not constitute thought experiments. Some thought experiments commence with jarring counterfactuals and unfold without introducing any new empirical evidence. A good example is that of Einstein's supposing he was traveling in tandem with a beam of light. The virtue of this account stems from its internal logic. Indeed, the logic is so compelling that there is no need to repeat the experiment or seek greater approximation to the actual world, as one would in laboratory work. By exaggerating the dimensions of the relevant variables or creating a situation that is thoroughly fictitious, we avoid confusion as to whether the mind alone is conducting the experiment. Moreover, the aim of the thought experiment is not that of measurement or identification, as is true of most laboratory experiments. Rather, it is usually to establish a tendency, a pattern, or a law; in that sense, the experiment is more a demonstration than an open-ended inquiry.[12]

Hume was one of the first to conduct extensive thought experiments in what we now call *economics,* notably on the subject of money.[13] In "Of the Balance of Trade," for example, he supposes first that four-fifths of all money is annihilated overnight and then, in another experiment, that the quantity of money is increased fivefold overnight. In both cases, he spells out the consequences for prices given foreign trade and shows that the quantity of money would be restored to its original level (Hume 1777/1985, 311). Here, and in similar trains of thought, Hume employs unrealistic quantitative shifts in order to help the mind grasp that it is the relative proportion of money to prices, not the absolute quantity of money, that matters. In other essays and letters, Hume entertains the notion of the complete and sudden annihilation either of gold or paper money, arguing that the void would be immediately filled by silver, in the first case, and foreign specie, in the second (296, 320). Money thus abhors a vacuum as much as nature does. The implication of this, for Hume, is that the importance of money should be downplayed—it will flow on its own accord—and the importance of people and industry underscored: "A government has great reason to preserve with care its people and its manufactures. Its money, it may safely trust to the course of human affairs, without fear or jealousy" (326). Hume's facility with quantitative thought experiments speaks to a skill that he may well have obtained from the study of natural philosophy.

Hume's own *Treatise* was an explicit "attempt to introduce the experimental method of reasoning into moral subjects," as the title page made plain (Hume 1739–40/2000, 1). Barfoot proposes that Hume's exposure to experimental science at Edinburgh motivated his many paeans to experimental reasoning in his moral philosophy (Barfoot 1990, 167). Hume's two years in France, first at Reims and then at La Flèche (1735–37), precisely when he wrote much of the *Treatise,* may have influenced him as well. One leading savant at Reims, Noël-Antoine Pluche, provided Hume with access to his library. Pluche was engaged in composing a multivolume treatise, *Le spectacle de la nature* (1732–50), that, among other things, mapped out recent developments in electricity and magnetism. Pluche and many of the scholars Hume would have encountered at La Flèche were Jesuits, who at the time were "leaders in experimental physics" and constituted "approximately half of the active electrical experimenters" (Hankins 1985, 55). In that respect, Hume probably fared better in his studies abroad than did Adam Smith, who found his six years at Oxford uninspiring (see Mossner 1980, 102; and Ross 1995, 70).

The first volume of the *Essays* that Hume edited for the Philosophical Society of Edinburgh contains Ebenezer McFait's "Observations on Thunder and Electricity," which refers to Benjamin Franklin's celebrated experiments on electricity (Philosophical Society of Edinburgh 1754–71, 1:209–18). In 1759, Hume met Franklin while the American was visiting Edinburgh and, in 1762, invited him to submit an essay on lightning rods for publication by the society. Notwithstanding his respect for Franklin, Hume may have had an additional motive. Lightning rods had recently become a politically charged subject, insofar as George III had endorsed the English recommendation of blunted rods, in opposition to Franklin's proposal for pointed ones (see Cohen 1990, chap. 8). Franklin's account of the lightning rod explicitly omits "the philosophical reasons and experiments on which this practice is founded; for they are many, and . . . are already known to most of the learned throughout Europe" (Philosophical Society of Edinburgh 1754–71, 3:132). This strongly suggests that Franklin had found Hume well versed on the subject in 1759, or he would not have made such a potentially embarrassing remark. In the third volume, there is also an article by a Dr. Austin describing the successful use of electric shock treatment, in 1764, on an adolescent girl suffering from an obstruction of her menses (Philosophical Society of Edinburgh 1754–71, 3:116–19).

There is, alas, no concrete evidence as to when Hume may have first learned about the theory of electricity, but the likelihood is high that it was before 1750. Hume's university courses may well have included some electrical experiments, and, in his travels in France during the mid-1730s, he may have seen some demonstrations by the Jesuits. In the *Dialogues concerning Natural Religion* (1779), there is a reference to electricity as one type of source for the motion of material bodies (Hume 1779/1947, 182). Hume completed the first draft of this controversial work in 1750–51, suggesting that he had already reflected on the role of electricity before then. Other members of the Edinburgh Philosophical Society, notably James Russell and Patrick Brydone, had a keen interest in electricity, an interest stemming back to the 1740s. There is also some evidence that Hume observed electrical displays first-hand, at the ancestral home of Sir James Steuart, before the latter's departure for France in 1746.[14]

The group at Edinburgh, especially Maclaurin, might first have read some of Franklin's work during the course of its 1743 debates on population growth. But, if not in 1743, then most likely soon after Franklin's theory of the electric fluid began to circulate in 1747. Hume's close friends Cullen, Maclaurin, Black, and Hutton all championed the doctrine of the ether as the site of chemical and electrical affinity. Moreover, they drew epistemological ammunition from Hume: "Ethereal speculation was seen by Hume as legitimate, a secular and therefore a preferable alternative to any reliance on a realm of divine causation in nature" (Christie 1981, 88). There is much to support the claim that Hume knew and valued the doctrine of the ether and its ramifications for the study of subtle fluids.

Deborah Redman has provided a detailed study of the contents of the *Gentleman's Magazine* for the years 1731–59. The *Gentleman's Magazine* was the leading periodical in Britain at the time and read by Hume and Smith, although whether regularly it is impossible to tell. Redman claims: "Even the most casual perusal of the journal in these years confirms what an integral role science played in eighteenth-century society and how important natural philosophy was for moral philosophy and vice versa" (Redman 1997, 361–62). In addition to numerous articles on medicine and astronomy, there were, from

1746 to 1751, some two dozen articles on electricity, including entries on electric shocks (1750–51) and on the medical uses of electricity. As one entry in 1745 made plain: "From the year 1743, they discover'd phenomena so surprising as to awaken the indolent curiosity of the public, the ladies and the people of quality, who never regard natural philosophy but when it works miracles. Electricity became the subject in vogue, princes were willing to see this new fire" (quoted in Schaffer 1983, 6). Indeed, there were few other entries on the physical sciences, apart from some articles on optics, meteorology, and geophysics (see Redman 1997, 368–71). Electricity clearly stood out as the most interesting subject in that five-year period. All this lends weight to my conjecture that Hume would have known about electrical experiments long before he edited McFait's article on the subject in 1754.

MONEY AND THE ELECTRIC FLUID

Hume may also have thought of money in terms of electricity, insofar as he treated money as a fluid during the very decade in which the doctrine of subtle fluids was ascendant. Of all the subtle fluids ca. 1740, electricity was the one that received the most investigation and popular acclaim. Hume issued "Of Money" in 1752. We know, from his correspondence with James Oswald, that the essay was drafted by October 1750 and that Oswald persuaded Hume to reconsider the impotency of money (see Rotwein 1970, 190–99). As Hume replied in November 1750: "The additional stock of money may, in this interval, so increase the people and industry, as to enable them to retain their money. Here I am extremely pleased with your reasoning" (Rotwein 1970, 197). Thus, in late 1750, Hume accepted Oswald's point that the quantity of money mattered and, presumably, changed "Of Money" quite substantially as a result. Hume went beyond Oswald's insights, however, in that he focused on the intensification of the supply of labor as the operative step. And we know from Hume's references to an "etherial active fluid" in the *Enquiry*, the hearsay evidence of James Steuart, and the popularity of Franklin's work, Hume was most likely aware of the new discoveries on electricity by the time he revised "Of Money."

As is well-known to historians of economics, Hume often depicted money as a kind of fluid and certainly heaped scorn on the fetish for bullion. To the best of my knowledge, the analogies of oil and blood were used once each (Hume 1777/1985, 281, 294); however, the use of water imagery, particularly the property of equilibration, was of considerable importance in his analysis of what we now label the *specie-flow mechanism*. Jacob Vanderlint had already identified this process in his *Money Answers All Things* (1734) and drawn a direct analogy to the ebb and flow of the tides (Vanderlint 1734/1914, 49). Hume also uses ocean imagery and the general observation that water always seeks its own level to reinforce the claim that money would readily flow from one country to another were trade unimpeded. There can be no doubt that Hume utilized some of the properties of water to make sense of the distribution of money.

Nevertheless, I believe that the more operative set of analogies involved fluids in general and the electric fluid in particular. An unanticipated increase in the money stock had the ability to engender real economic growth. Here we see that Hume did not simply evince the more neutral attribute of water coating, but not transforming, objects. While he may not have deliberately adopted analogies to the electric fluid, I believe that he conceived of money as having certain properties that other fluids, especially ordinary water, did not. These notions may have been sufficiently commonplace to Hume and his contemporaries not to require explicit mention. The fact that he could employ different analogues (oil and blood) and, in some cases, speak of money as simply a "fluid" suggests all the more that he had few qualms about adopting whatever imagery suited his purpose (Hume 1777/1985, 320–31).[15]

For one thing, money could be conserved (since it did not evaporate) and even condensed, as in a bank (or Leyden jar).[16] For another, it had the same tendency to ubiquity that was ascribed to electricity. Hume emphasized the importance of the "universal diffusion and circulation" of money. As a leading exponent of the new commercial order, he envisioned a time when money would coat all transactions, much as the electric fluid was thought to cover the entire surface of all bodies. He depicted the seeping of money into every nook and cranny of the economy, celebrating the day when "no hand is entirely empty of it" (Hume 1777/1985, 294). In this respect, the diffusion and circulation of money resembled well the universal diffusion of the electric fluid. By about 1740, it was established that every substance was capable of electrification, which prompted many eighteenth-century physicists to speculate that the electric fluid was the most fundamental stuff, filling the interstices of atomic particles and, thus, permeating all matter. Leading savants such as Stephen Gray and Edmond Halley conjectured that the electric ether was a "universal medium" and the seat of all powers (Schaffer 1983, 7).

Moreover, some French philosophers, such as Dufay, Maupertuis, and Buffon, conjectured that electricity might be the source of life itself, an idea that would have resonated well with Hume's secular leanings. This would make sense of Hume's attribution of vital properties to money: "In every kingdom, into which money begins to flow in greater abundance than formerly, every thing takes a new face: labour and industry gain life" (Hume 1777/1985, 286). This image of money inducing a sudden surge of vitality, followed by a period of dissipation and the return to a normal level of labor input, does not make as much sense if the analogy to ordinary water is assumed. Hume also proposes that the best policy for the magistrate is to keep the money stock increasing "because, by that means, he keeps alive a spirit of industry in the nation" (288).

It is important to step back and note just how unclear we are even to this day about the precise path by which this process of an unanticipated inflation having real effects comes to pass.[17] Since Nicholas Copernicus and Jean Bodin, we have countless observations of the correlation between the money supply and the price level, but how these are actually connected in terms of the decisions and actions of individuals, and how there might be an intervening stage that has real effects on economic productivity, still eludes our full understanding. One leading expert, David Laidler, has suggested that the theory remains controversial precisely because the causal chain could be the opposite of what is supposed: prices rise, and money matches the increase in the price level. This uncertainty stems partly from the fact that we cannot experiment directly on the economy and, thus, isolate the relevant variables with complete certainty.[18]

Numerous commentators, including Laidler, have praised Hume for giving a detailed physical description of the quantity theory of money, specifically the process by which output and, subsequently, prices are increased by a sudden and unanticipated rise in the money supply (see Duke 1979). But even Hume does not provide a full description. How, precisely, do people perceive a change in real balances (a heavier sensation in their pockets?), and why does this intensify the supply of labor, even in the case, as Morris Perlman (1987, 282) has persuasively argued, of full employment? The very first step that merchants take after finding that they have an excess of gold and silver is to invest in manufacturing. We find, then, that "the manufacturer [is] more diligent and skilful, and even the farmer follows his plough with greater alacrity and attention" (Hume 1777/1985, 286). The reason for this intensification of the work effort is simple: "Their industry [is] only whetted by so much new gain" (287).

Yet it is still unclear how the new injection of money results in "quicken[ing] the diligence of every individual, before it encrease the price of labour" (Hume 1777/1985, 287). After all, only merchants have the virtue of frugality. And the normal tendency among laborers, for Hume at least, is toward indolence and apathy (299). The sudden arrival of new money works its first effect almost instantaneously, as though the causal chain is strictly physical, without any conscious reflection on the part of the people concerned. In this case, not just the merchants, but all who labor are stimulated to intensify their work effort, almost as though charged by some physical force. The effect, however, is not long-lived. Vendors quickly grasp that there is an excess demand for goods, prices catch up, and the work effort returns to its prior level. As has often been noted, however, the time that elapses from the start of this process to its finish is left unspecified. In short, the entire process, the chain of mental and physical inferences and actions, is, for someone such as Hume, under-specified.[19]

Perhaps, having witnessed some popular electrical parlor games (see chapter 2), Hume accepted more readily the idea that money could increase productivity. Its powers too could be conducted from one body to another, much as a person could absorb the electric fluid and charge up others around him before returning to a neutral state. Recall that, for Hume, causation works indiscriminately between the physical and the mental, or the natural and the moral, in Hume's terminology. In that sense, Hume's celebrated account of the quantity theory of money allows for the physical or vitalistic properties of the monetary fluid as much as it does for the mental ones.

In sum, Hume may have been prompted to explore such properties as conservation, diffusion, capacity, condensability, ubiquity, and vitality because of his knowledge of recent investigations on the electric fluid. This would help account for the remarkably original character of his monetary analysis. Previous thinkers had compared money to water or blood, but Hume was unprecedented in his command of monetary fluctuations, the rapid flow of money from one region to another, and the return to a state of equilibrium both locally and globally. Moreover, he drew a firm distinction between the outcomes of anticipated and unanticipated growths in the money supply. In the latter case, money has the power to make significant, if short-lived, changes to patterns of work and exchange. Perhaps the now commonplace term *monetary shock* has more literal connotations than we have recognized.[20]

No one disputes that Hume makes extensive use of fluid analogies in his efforts to understand money. I have here pointed out that he did so at a time when experimental natural philosophers were steeped in the study of subtle fluids, especially electricity. Not everything was conceived of as a fluid, but certainly many more things were discovered to be fluid-like than had previously been thought. It is my belief that Hume was influenced by the widespread contemporary effort to treat common but mysterious substances as if they were fluids and that this in turn accounts for his ideas on the behavior of money. The fact that the theory of money is also the area in which Hume most often conducts thought experiments may be additional evidence that the broader context of his political economy was that of natural philosophy. Hume was remarkably insightful on the subject of money, both in noting what is now called the *nominal/real distinction* and in tracing out various consequences of monetary adjustments. Both his thought experiments on money and his general conception of money as a fluid enabled Hume to climb several rungs up the ladder of monetary abstraction (see Gatch 1996).

NATURAL HISTORY

Another strong analogy drawn between the physical and the moral realms can be found in Hume's appeals to cyclic processes. The world, like an individual, passes, "by corruption or dissolution, from one state or order to another" (Hume 1777/1985, 377). All creatures—man, animal, and vegetable—partake in these cycles of advancement and decline. These appeals to growth and decay are also found when Hume appeals to the moral realm. As Eugene Rotwein noted: "Hume's economic thought takes the form of a natural history of the 'rise and progress of commerce'" (Rotwein 1970, xxxii). Wealth, the arts, and the sciences flourish in one place, then decline, only to flourish elsewhere (Hume 1777/1985, 378; see also Skinner 1993b, 244). Moreover, the cycles reinforce one another. Hume claims that one cannot encounter flourishing arts and sciences without also encountering flourishing commerce—and a liberal government as well. And commerce comes only if the various activities—agriculture, manufacturing, and trade—coincide in one region. Furthermore, each region's apotheosis is short-lived: "*When the arts and sciences come to perfection in any state, from that moment they naturally, or rather necessarily decline, and seldom or never revive in that nation, where they formerly flourished*" (Hume 1777/1985, 135). The justification of this claim is based on an analogy to plants: "The arts and sciences, like some plants, require a fresh soil; and however rich the land may be, and however you may recruit it by art or care, it will never, when once exhausted, produce any thing that is perfect or finished in the kind" (137).

In his analysis of money and prices, Hume outlines a mechanism that allows a more rapid flow of wealth from one region to another. When domestic labor becomes too costly, manufacturers move en masse like a flock of birds, almost "flying" to other countries where wages are lower, until they "are again banished by the same causes" (Hume 1777/1985, 283–84). Money readily adjusts in order to restore domestic prices to their "natural level." It is again only over larger chunks of time—centuries, given his analysis of population growth—

that, for Hume, wealth really intensifies or diminishes in a given region. There are, in short, built-in checks to the tendency for wages and prices to rise, just as urban centers tend to reach a saturation point (which, Hume suggests, is the case for London [448]).

There is also a natural progression by which the flourishing of one region is followed by that of another. In a well-known letter to Lord Kames of 1758, Hume wrote: "It was never surely the intention of Providence, that any one nation should be a monopolizer of wealth: and the growth of all bodies, artificial as well as natural, is stopped by internal causes, derived from their enormous size and greatness. Great empires, great cities, great commerce, all of them receive a check, not from accidental events, but necessary principles" (Rotwein 1970, 201). Note that Hume readily equates the natural and the artificial, implicitly comparing animals and nations.

The only question left to resolve is the direction in which human society is headed.[21] Hume maintains that "the universe, like an animal body, had a natural progress from infancy to old age" but that, since it is uncertain whether we have reached middle age as yet, "we cannot thence presuppose any decay in human nature" (Hume 1777/1985, 378). He also argues at length that population now exceeds that of ancient times and that this is correlated with happier and more virtuous conditions (382), which he sees in the increase of gallantry, in the decline of political rivalry and vicious slaughter, and in the diminution of slavery and petty tyranny (see Schabas 1994b, 128–32). And he puts much faith in the growth of trade and commerce, which he asserts despite the paucity of records to that effect. Our ignorance, Hume conjectures, favors optimism: "It is not fully known, what degree of refinement, either in virtue or vice, human nature is susceptible of; nor what may be expected of mankind from any great revolution in their education, customs, or principles" (Hume 1777/1985, 87–88).

These shifts take place over long stretches of time, over centuries, and are, thus, imperceptible to any contemporary analysis (Hume 1777/1985, 378). Hume suggests that a three-thousand-year written record is all too brief "to fix many general truths in politics" (87). But there is an inevitability to the growth and eventual decay of all human activities. Even human nature, while stable and robust in certain respects, undergoes an evolution over time. Certain traits, such as politeness or scientific curiosity, are induced by favorable material circumstances. Trust is more deeply entrenched, as is evident in the increase in the number of cartels and the use of fiduciary money (406).

Hume may well have favored cyclic accounts because of his appreciation for Stoic thinking, but there is in his thought also a direction to human history, suggesting that he was already favorably disposed toward the evolutionary schemes that were emerging in his day. In the library attached to Steuart's course at the University of Edinburgh are nine books on the theory of the earth's creation as well as a much larger collection of works on natural history. Whether Hume read some or all of them is still open to question, but certainly by the 1740s he had reflected on such subjects when drafting his *Dialogues concerning Natural Religion*.[22] Roy Porter has argued that theories of the origins of the earth gained in popularity in the late seventeenth century and the early eighteenth because of their secularizing and even political implications. Moreover, earth histories were formulated partly to reposition man in nature, as in the case of Locke's state of nature (Porter 1979). Hume's *Dialogues* easily fits into Porter's account, although he failed to cite that particular work.

The 1740s and 1750s marked a watershed in the development of theories of generation and natural history more generally. Charles Bonnet's work on aphids (1740) and Abraham Tremblay's study of hydra (1744), by lending evidence to spontaneous generation, had intensified debates on the origin and nature of life. Both Julien La Mettrie and Dénis Diderot embraced Tremblay's mechanistic theories, their work in turn inspiring Pierre de Maupertuis and the Comte de Buffon, by the late 1740s, to broach evolutionary hypotheses. Evolutionary ideas circulated widely with the publication of the first two volumes of Buffon's *Histoire naturelle* in 1749—and possibly as early as 1744, when the first part of volume 1 was presented before the Academie Royale des Sciences.[23]

It is hard to believe that Hume, who had such a strong affinity with the secular French philosophers and who was at the center of Enlightenment debates, did not hear about these ideas at least in their simpler versions.[24] Adam Smith, in a 1776 letter to the *Edinburgh Review,* explicitly addressed the importance of Buffon's work on generation, conveying the impression that it was already well-known—and controversial—for its materialist and atheistic implications (see Wood 1989, 99–100). Alas, there is no concrete evidence that Hume read Buffon until a decade later, when the two met in Parisian salons during the 1760s. But Buffon's gift to Hume of a personal copy of his *Histoire naturelle* suggests that they conversed on scientific issues (see Mossner 1980, 480). Hume also expressed a distinct liking for the French naturalist and sounds distinctly Buffonian in the following observation: "The continual and rapid motion of matter, the violent revolutions with which every part is agitated, the changes remarked in the heavens, the plain traces as well as tradition of an universal deluge, or general convulsion of the elements; all these prove strongly the mortality of this fabric of the world" (Hume 1777/1985, 377).

James Hutton, arguably more than any other Enlightenment figure, stretched our estimate of geologic history (see Laudan 1987). In defiance of the biblical account, Hutton proposed that the earth's geology displayed "no vestige of a beginning,—no prospect of an end" (Hutton 1788, 304). Hutton was a deist, and this might well have been partly why he and Hume formed a bond starting in the late 1740s. It is difficult to know whether Hutton held these views at that point in time since his geologic findings came later, in the 1760s. Moreover, his views were not widely known until he published the *Theory of the Earth* in 1795, two years before he died. He had studied medicine at Leiden, where the Boerhaavian school was still in ascendance, and then befriended Black, Hume, and Smith. A direct influence is, thus, difficult to establish, although Hume's appeals to eternal cycles resonate with Hutton's geology. Hutton aside, if Hume was aware of new currents in French natural history, this would suffice to account for his emphasis on economic evolution and the relatively unprecedented temporal element in his essays.

Hume may aptly be grouped with the Scottish historical school and certainly evinced a general love for history, as manifest in the lengthy and insightful *History of England* and in the 1741 essay "Of the Study of History."[25] But the specific evolutionary tenor of his economics suggests a more localized source. For Hume was, in many ways, the first to compare and contrast economies by the century, if not over broader stretches of time. In "Of Money," the most prevalent temporal interval is three centuries (Hume 1777/1985, 281, 289, 292, 294). Hume draws comparisons, not only between the English and the German economies, but also between the European and the Chinese and refers back to ancient Rome in three passages (282, 285, 294). Thus, when he points to the "happy concurrence of causes in human affairs, which checks the growth of trade and riches" and prevents any one country from prolonged dominance (283), he could have only several centuries in mind, if not just the thousand-plus years that separated the fall of Rome and the point early in the eighteenth century at which the wealth of Western Europe surpassed that of Rome. Elsewhere, he compares the present level of industry in Britain to its state there two centuries ago (328). No mercantilist or Physiocratic writer has the same temporal sweep. I would submit that Enlightenment naturalists such as Buffon and Hutton served to awaken Hume's mind to the *longue durée*.

Human action takes center stage for Hume, but the backdrop throughout is that of a natural and orderly world, one that is configured so as to foster human prosperity. On reading Hume's many essays, say "Of the Rise and Progress of the Arts and Sciences" or "Of the Populousness of Ancient Nations," one is struck by his command of an immense sweep of global history. Hume unfolds a tale of passions and human frailty on a vast temporal and geographic scale, drawing numerous lessons from the rise and fall of the empires of Greece, Rome, and Spain, among others. Wealth and virtue ebb and flow in accordance with a complicated morality tale. Hume even peers well into the future, conjecturing in "Of Public Credit" that, in five hundred years, servants and masters will have changed stations (Hume 1777/1985, 357).

CONCLUSION

I have now examined certain aspects of Hume's economic thought that appear to be influenced by ideas drawn from natural philosophy. Both his monetary theory and his historical approach to economic development were part of a much broader framework whereby Hume was attempting to understand the moral realm and grapple with the more immediate issue of political stability. My interpretation in the more specific sense has been established by noting significant correlations. I have no evidence pointing definitively to causal connections, but I do think that there is enough circumstantial evidence to make my case. Moreover, Hume consistently approached human activity, economic and otherwise, with an eye to physical nature. There are a sufficient number of metaphysical and epistemological points of similarity between his economics and contemporary natural philosophy to suggest that Hume regarded economic processes as part and parcel of nature.

Hume's writings are so dense and weighty that it would be rash to generalize from such a brief overview. Moreover, Hume would be the first to say that the mere conjunction of two events does not legitimate the assertion of a causal link between them. Yet I trust that the evidence drawn from his life and writings that I have presented in this chapter warrants the view that Hume's moral philosophy and certain aspects of his economic thinking were infused with ideas and methods drawn from natural philosophy. Hence his conjecture that there exists "a kind of pre-established harmony" (Hume 1748/2000, 44) between our minds and nature that serves as the single source from which all knowledge, including knowledge of economic phenomena, emanates.

CHAPTER FIVE

Smith's Debts to Nature

Smith's formulation is that nature did not leave it to man's feeble reason to discover that and how he ought to preserve himself, but gave him sharp appetites for the means to his survival.

—Joseph Cropsey, "Adam Smith and
Political Philosophy"

Adam Smith (1723–90) was *the* towering figure of Enlightenment political economy, a stature that he attained in his own lifetime, much as Newton had become preeminent in his.[1] The secondary literature on Smith is enough to sink a small boat; there are more than one thousand books and journal articles. Unfortunately, or perhaps fortunately, most of these are Whiggish efforts to show just how modern Smith was in his economic analysis. As Jacob Viner once remarked: "An economist must have peculiar theories indeed who cannot quote from the *Wealth of Nations* to support his special purposes" (Viner 1991, 92). Such versatility has lent itself to considerable abuse among modern economists in search of venerable ancestry.

Smith did not, however, view himself as an economist. Indeed, the term *economist* was not in common usage in English at the time.[2] Smith was a professor of logic and moral philosophy with a wide range of interests, including jurisprudence, natural philosophy, and rhetoric and belles lettres. Although we have reason to believe that he wrote voluminously, he shepherded only two books into print, *The Theory of Moral Sentiments* (1759) and *The Wealth of Nations* (1776).[3] Both were revised substantially after their initial publication. Smith lived to see the sixth edition of his first book appear just weeks before he died, and the sixth edition of the second appeared the following year. For Smith, *The Theory of Moral Sentiments* was the more significant of the two, but it waned from the philosophical firmament after Kant and, while increasingly appreciated in recent years, has yet to regain canonical standing.[4]

As numerous scholars have argued, natural philosophy and moral philosophy were closely joined in the Scottish Enlightenment, arguably more so than in France or elsewhere in Northern Europe (see Emerson 1990; and Wood 1990). Smith, needless to say, was a shining example of this felicitous union. He took up Newton's challenge to extend the rules of reasoning into uncharted land and was regarded by others, such as John Millar and Thomas Pownall, as "the Newton of the moral sciences" (see Redman 1997, 208–15). My objective here is to adumbrate Smith's indebtedness to the concepts rather than the methods of early modern natural philosophy. Some of the many topics in Smith's economics where nature rears its head are labor, markets, and the pursuit of wealth. Before explicating these many points of intersection, however, I will spell out the extent of Smith's scientific knowledge as well as his use of the terms *nature* and *natural.*

SMITH'S KNOWLEDGE OF NATURAL PHILOSOPHY

We know that Adam Smith was well versed in the classics, mathematics, and natural philosophy, a grounding that he received first in Scotland, at his local school and at the University of Glasgow (1737–40), and then, to a lesser degree, at Oxford (1740–46). His most influential teacher was Francis Hutcheson, who served as professor of moral philosophy at Glasgow, but who also had a solid grounding in logic and natural philosophy, including Newtonian physics. Hutcheson was known for promoting the methods of Newton in other fields and is credited with inspiring Smith to do just that in moral philosophy. One of the more memorable courses that Smith took from Hutcheson was on the subject of "Pneumaticks," by which was meant the "science of spirits or spiritual beings" (Ross 1995, 43). What little evidence we have suggests that Hutcheson treated the subject as speculative physics (see Moore and Silverthorne 1983). He covered metaphysical questions about ethereal beings as well as the new physics of airs and other elastic fluids. Before Hutcheson, Gershom Carmichael had taught the course, and, in its earlier incarnation, it had a strong allegiance to natural theology. But Hutcheson toned down the religious content and adopted much the same sense of a providential order that we find in Smith.

Smith also took courses from Robert Dick on experimental physics, a subject that was much emphasized in the Glasgow curriculum. According to Ian

Simpson Ross, the Scottish universities were keen to be "modern" and, thus, as much as possible kept abreast of recent innovations in experimental natural philosophy, unlike the more traditional Oxford.[5] At Glasgow, an additional fee of three shillings per session was levied on each student to fund the purchase of equipment. While some equipment had been purchased in the late seventeenth century, the collection was mostly developed in the early to mid-eighteenth century. In 1726, the university purchased twenty-eight pounds worth of scientific apparatus built by Francis Hauksbee, the well-known instrumentmaker for the Royal Society. Following this, annual expenditures on laboratory equipment ranged as high as 350 pounds, the amount spent in 1763 when Smith, as vice rector, authorized a new laboratory for Joseph Black.

In contrast to his three stimulating years at the University of Glasgow, Smith's sojourn at Oxford was uninspiring. After Newton's death in 1727, the pursuit of mixed mathematics declined in England, particularly in comparison to France. Oxford was not a place for serious study, and most students treated it more as a drinking society than as an academy. Although his scholarship at Oxford was intended to train him for the Church of England, Smith soon gathered that he was not so inclined, and, in 1749, he resigned his status as a Snell Exhibitioner. We do know that he read intensively during these years and continued his studies of numerous subjects, including civil law, classics, and possibly physics, where the focus was primarily on Cartesian mechanics, with some coverage of Newton and 'sGravesande. Insofar as the evidence supports a youthful drafting of the essay "The History of Astronomy," which covers Cartesian metaphysics in some detail, there is reason to think that he did continue his study of physics. He probably read Hume's *Treatise of Human Nature* soon after it appeared (1739–40) and Bishop Berkeley's works, particularly his theory of vision (see Ross 1995, 77).

After his student days, Smith forged friendships with the intelligentsia of Scotland. The most intense and enduring of these was with David Hume, whom Smith met ca. 1750 in Edinburgh, and it lasted right up until Hume's death in 1776. Because there are so few extant letters between the two, we can only begin to imagine their conversations on philosophy and political economy. Their mutual esteem, as well as frank criticism of one another's ideas, shines through in the letters that have survived. Another friendship forged at this point in time was with William Cullen, who found in Smith "a like-minded ally" (Donovan 1975, 72). Cullen had been appointed as a professor to the University of Glasgow in 1751, the same year as Smith, and then moved to the University of Edinburgh in 1755, where he is largely credited for turning the medical school into the most prestigious one in Europe. He found the school run by disciples of Hermann Boerhaave, the great Leiden physician, who treated the body as a hydraulic machine completely severed from the soul. Cullen, along with Robert Whytt and Alexander Monro, transformed the study of medicine by treating the body as essentially neurological. While such an approach may have had no greater therapeutic value, it meshed well with the associationist psychology of Locke and Hume as well as with the drift toward vitalism and away from mechanistic thinking. There was also, as we will see, a reciprocal influence between Cullen and Smith. Cullen emphasized the nervous system in part because he was convinced by Smith that sympathy was the most fundamental of human sentiments. Likewise, Smith drew on Cullen's neurology when he depicted sympathy as essentially a physiological reaction to the plight of others (see Lawrence 1979; Forget 2003).

Cullen was a Newtonian when it came to method but not when it came to metaphysics (see Schofield 1970, 206–9, 218). He found his teacher, John Desaguliers, too rigid and faithful to the great Sir Isaac. Moreover, Cullen cultivated a strong fascination with the concept of the ether as the seat of natural powers.[6] This in turn led him to endorse Stephen Hales's doctrine of subtle fluids, as an improvement over the mechanical philosophy of the seventeenth century. Cullen appears to have shared Hales's suspicions, noting that the mechanical philosophy "never could, nor ever can be applied to any great extent in explaining the animal economy" (quoted in Schofield 1970, 206).

The idea that solid bodies could absorb and hold air was first broached by Hales in the 1720s. Although Smith did not know Hales personally, he did, we know, admire his work and, presumably, would, thus, have followed the development of these ideas in the hands of Cullen and Joseph Black, who succeeded Cullen at Glasgow (see Donovan 1976). In 1754, Black was the first to isolate a new "fixed air," and he communicated his results in correspondence with Cullen. The official publication came two years later and constituted a major step toward undermining the ancient authority of the doctrine of four elements. Fixed air, or carbon dioxide, as we now know it, was released on heating magnesia alba and found to be highly soluble in water (seltzer water). The air is, thus, "fixed" in both solids and liquids—in what subsequent experiments showed could be quite substantial amounts. Cullen used Hales's ideas to investigate the properties of evaporation, and both Cullen and Black extended this process in their respective theories of heat as a substance, a caloric fluid. Black also developed the concept of latent heat, which, along with that of fixed air, ushered in the so-called Chemical Revolution of the late eighteenth century (see Golinski 1992).

Both Black and James Hutton became good friends with Smith in the middle decades of the eighteenth century. After Hume passed away in 1776, they accepted Smith's invitation to serve as his literary executors and were present when Smith died. In 1759, Smith befriended Benjamin Franklin. While the extant record indicates a shared concern for political unrest in America, there is reason to believe that Smith, like Hume, also appreciated Franklin's scientific work. In the essay "Of the External Senses," Smith cites Franklin's experiments on the propagation of sound (Smith 1795/1980, 147).[7] And his library included Franklin's famous letters on electricity (see Mizuta 1967/2000). Following his appointment to the Royal Society in 1773, Smith would have rubbed shoulders with the leading English natural philosophers of the day, notably Henry Cavendish. All this suggests that Smith was au courant with recent developments in science.

Smith's library contained a wide selection of books on natural philosophy. Included were the writings of such prominent mathematicians and mechanical philosophers as Galileo, Christian Huygens, Robert Hooke, Pierre Gassendi, Newton, and Colin Maclaurin, those of the prominent chemists Pierre-Joseph Macquer and Joseph Priestley, and those of the prominent botanists and zoologists Marcello Malpighi, Lazzaro Spallanzani, Carl Linnaeus, Charles Bonnet, and Georges-Louis Leclerc de Buffon (see Mizuta 1967/ 2000). Smith also had an extensive set of volumes of the *Philosophical Transactions of the Royal Society,* one of the leading periodicals for scientific discoveries. In 1780, Smith wrote to a Danish friend that, for the past six years, he had, in addition to revising *The Wealth of Nations,* studied botany and "some other sciences to which I had never given much attention before" (Ross 1995, 227). There is reason to believe that Smith sustained an interest in natural philosophy over the course of his career.

Lawrence Dickey has suggested that, for *The Theory of Moral Sentiments, nature* was a "scene" term, whereas, for *The Wealth of Nations,* it had become a term of an "agent" (see Dickey 1986, 604). Indeed, both works are so peppered with the words *nature* and *natural* that more than one scholar has attempted to catalog Smith's use of them. Edward Puro (1992) has identified eight different usages of the word *natural* in *The Wealth of Nations* alone. And Charles Griswold (1999), looking to the moral but not the jurisprudential works, has found seven categories that encompass both the natural and nature.[8]

I am least concerned here with those situations in which Smith uses the term *natural* to mean something that is normal or commonplace, for example, a reference to merchants "naturally" conspiring to raise prices. More significant for my purposes is his use of appeals to nature as a type of essentialism.

Mankind, for example, is "naturally sympathetic" (Smith 1790/1976, 21), a fact that is critical for the preservation of human society. Like Aristotle's, Smith's essentialism covers the entire gamut of objects, not just human behavior. In this sense, nature provides us with patterns; what is observed to be regular must originate in nature. Hence, Smith refers often to "human nature" or to "the nature of things" in general. Another distinct way in which Smith makes use of the term *natural* is to contrast that which is natural with that which is artificial, or that which comes about by human contrivance. This is of critical importance in his appeals to "natural liberty" and possibly "natural price." As T. D. Campbell put it: "Nature meant for Smith the humanly unhindered or unobstructed, and this more amply means what is not confounded by the mis-placed interventions of human reason: letting nature take its course, letting men do as they are instinctively prompted to do" (Campbell 1975, 148).

At some level, however, this distinction does not make sense. If human na-ture is part of nature, then within nature's compass must fall all that emanates from human nature. Even if one could isolate the set of "misplaced interven-tions of human reason," one would need to explain their source and trace them back to some aspect of human nature. Smith emphasized the primacy of in-stinct over reason, thereby reminding us of our bonds with other animals and, thus, with natural history, but he was also inclined to view nature as the sum total of objects in the universe. For Griswold, this is Smith's most significant use of the term, and the one that most embodies his Stoic beliefs (Griswold 1999, 316). Nevertheless, Smith broke away from the Stoic conception of na-ture, in which man is seamlessly joined with the rest of the natural world (Gris-wold 1999, 321–24). He also took issue with the Stoics' refusal to privilege man over other life-forms. Clearly, for Smith, humans are, not only central to the operations of nature, but also "destined . . . to be the governing animal in this little world" (Smith 1795/1980, 136). We are, thus, part of nature even though the laws that govern our behavior and enable us to rule the natural kingdom are of a different kind.

Smith attributes numerous characteristics to nature. For one, nature is or-derly because it was designed by a deity. Hence, there are certain well-behaved phenomena in the world, such as the orbits of the planets, that are part of a di-vine plan. Smith is convinced that there are patterns in the world, or at least that it is our mission as philosophers to discern such patterns at the level at which they are presented to us. On numerous occasions, he refers to the chains that bind together the phenomena of nature. Nature is also personified.[9] "She has provided for all her children . . . parental care" (Smith 1790/1976, 52, 148). She does this with a "peculiar and darling care" and "exhorts mankind to acts of beneficence" (86). "Nature" is also an instructor, who "teaches us to hope" in accordance with a "doctrine" (91, 53). In keeping with her pedagogical role, she rewards and punishes according to certain laws, which are "dictated" to us by her (82).

Nature is also a book, whose "Author" evinces perfect wisdom, oeconomy, and proportion. As a result, "Nature" has designed man to accord with her intentions (Smith 1795/1980, 163). These are primarily the "happiness and perfection of the species" (Smith 1790/1976, 105), a notion that includes our survival and reproduction: "Thus self-preservation, and the propagation of the species, are the great ends which Nature seems to have proposed in the formation of all animals [including mankind]" (77; see also 87). For example, parental regard is much more passionate than filial piety. Moreover, "Nature" has not "abandoned us entirely to the delusions of self-love"; rather, she has endowed us with "an original desire to please" others (159, 116). She has also designed us to do some of her work: "Thus man is by Nature directed to correct, in some measure, that distribution of things which she herself would otherwise have made" (168). There is an implicit pedagogical aspect to the rules that "Nature" has given us in that we must also learn to formulate rules of our own. But both her rules and ours are unilateral in their objectives: "Both are calculated to promote the same great end, the order of the world, and the perfection and happiness of human nature" (168). This helps explain why a policy of laissez-faire is never sufficient. Nature cannot be left to take its own course; rather, it must be guided by human understanding and the rules that that understanding has devised.

Clearly, *nature* is a polyvalent term for Smith, carrying considerable metaphoric baggage. This is no doubt true for many, if not all, eighteenth-century savants, but there is in Smith at least a clear sense in which the laws governing the moral realm, including economic phenomena, are entangled with the laws governing physical nature. Moreover, as we will see, the junctures may be substantive and not just metaphoric.

SMITH ON THE HISTORY OF SCIENCE

Smith's posthumous "History of Astronomy" reflects, not just a love for science, but a fascination with the motivations that prompt its pursuit in the first place. We know that Smith specifically requested that the essay be published after his death and that he placed considerable pride in his command of the subject.[10] Black and Hutton, as his literary executors, noted that the essay was part of a larger, unfinished project, which might explain why Smith never ushered it into print despite its long genesis. As far as the essay itself is concerned, the last section on Newton was left unfinished, and there is no conclusion to balance the elaborate introduction. Published with the essay were two more posthumous contributions to the history of science, namely, "The History of Ancient Physics" and "The History of Ancient Logics and Metaphysics."

We have reason to believe that Smith first sketched "The History of Astronomy" soon after he left Oxford in the late 1740s and that the version that reached print was completed between 1752 and 1758, prior to the publication of *The Wealth of Nations* and *The Theory of Moral Sentiments* (see Hetherington 1983; and Cleaver 1989). There are clear indications of the influence of Hume's *Treatise,* such as references to "custom" and the significance of resemblance as the starting point for philosophical inquiry.[11] Oddly, Smith makes deprecatory remarks about chemistry, which suggests that he had yet to come under the influence of Cullen and Black (Smith 1795/1980, 46). Smith also bestows much praise on Descartes, which would have been appropriate in the 1740s, but not in the mid-1750s. In a letter to the *Edinburgh Review* of 1756, Smith accurately observes that the French had shifted their allegiance from Cartesian to Newtonian physics: "They seem now however to be pretty generally disengaged from the enchantment of that illusive philosophy" (Smith 1795/1980, 244). And, in a later version of the essay on astronomy, Smith cites several recent findings that enabled Newton's cosmology to supersede that of Descartes.[12] All this suggests that the essay did not undergo much revision after the early 1750s. Even in the 1773 letter to Hume, who was to serve as his literary executor, Smith notes that the essay covered the history of astronomy "down to the time of Descartes." Most likely, he had yet to complete the section on Newtonian physics or was all the more persuaded of its instrumentalist standing through the ongoing influence of Hume.[13]

"The History of Astronomy" provides us with the best glimpse into Smith's fascination with the natural sciences prior to his work on moral and economic philosophy. Nature presents us with phenomena, both wondrous and mundane, and the task of the philosopher is to find the principles that connect the phenomena, guided by analogy and resemblance. That there are such connecting principles Smith seems never to doubt, but he doubts very much that we can ever completely uncover them. The chains are analogous to those found in machinery, a favorite source of comparison for Smith. He would have had in mind a limited set of examples, such as a mechanical clock, the Newcomen engine, or the pianoforte, arguably the most intricate piece of machinery then in existence. Elsewhere in "The History of Astronomy," Smith refers to machinery used to produce the scenery in opera theaters, a comparison that conjures up an element of deception (Smith 1795/1980, 42).[14] In another essay, "Of the External Senses," Smith suggests that the imitative arts such as painting are but feeble representations of nature: painting "never has been able to equal the perspective of Nature, or to give to its productions that force and distinctness of relief and projection which Nature bestows upon hers" (Smith 1795/1980, 160). In that sense, "Nature" always outdoes what man can achieve, but there are at least approximations to her mysterious ways.

That the machines that Smith has in mind are for imitative purposes is apposite since the main point of "The History of Astronomy" is the development of what we would now call an *instrumentalist* position. Astronomers over the centuries have offered different theories or systems to make sense of the phenomena of the heavens, and, while one can see a gradual increase in the richness and accuracy of these systems, there is no definitive answer to be found among them. All are able to account for the phenomena in question, but accounting for phenomena is not necessarily the same thing as identifying underlying mechanisms. The Tychonic system helps underscore the possible range of such instruments. Moreover, Smith observes, Copernicus was forced to resort to the device of the (minor) epicycle despite his strong desire to eliminate it from astronomical systems.[15] The Copernican system's mechanical inconsistencies were, as Smith points out, fully compensated for by its empirical advantages, particularly its overall rendition of retrograde motion as well as of the restricted orbits of Venus and Mercury.

The Cartesian system was seen by Smith as a significant improvement, both for linking terrestrial and celestial phenomena and for providing a mechanical account, however fanciful. Descartes offered "a vast, an immense system, which joined together a greater number of the most discordant phaenomena of nature, than had been united by any other hypothesis" (Smith 1795/1980, 96). Smith elsewhere suggests that Descartes is as much an advance over Aristotle as Newton was over Descartes (244). Newton's theory is the most sophisticated of all, in that it partly accounts for the lunar orbit and can calculate the respective densities and distances of the planets. Newton's system is one "whose parts are all more strictly connected together than those of any other philosophical hypothesis" (104). Smith, however, does not present the Newtonian theory as the definitive word on the subject, only as the one to offer the most satisfactory account to date. He is careful to call it a hypothesis. What should be emphasized is his conviction that the chains that connect phenomena, including highly disparate phenomena, are what give Newton's account its epistemological strength. And, for Smith, the reason lies in the fact that nature is, indeed, law governed; there are "real chains" that "bind together her several operations" even if we may never know the correct or most fundamental chains (105).

Scholars have long identified the emphasis on concepts of nature in Smith's political economy (see, e.g., Skinner 1974; Campbell 1975). Vernard Foley has a book devoted to the subject that traces the Stoic and Cartesian components of Smith's naturalism (Foley 1976). More recently, Charles Clark provides detailed coverage of Smith's use of the natural law tradition, of arguments from natural theology, and of Smith's appeals to human nature as the foundation for an orderly moral realm (Clark 1992). Perhaps the most obvious example, one noted by numerous scholars, is Smith's figurative use of gravitational attraction in his treatment of market prices. He draws an explicit contrast between the observed price in the marketplace and the "natural price" to which "the prices of all commodities are continually gravitating" (Smith 1776/1976, 1:75; see also Fiori 2001). While the analogy cannot be pushed too far, it does hold for several features of Smith's schema. One feature is the attribution of market forces, which are by their very nature only inferred (from their effects on prices). Newton maintained the same about the occult force of gravity, that it could only be inferred (in this case from its effects on bodies). Another feature is that the market forces are central. That is, under the influence of these ever-present forces, a price will always tend toward its natural level, whether it is currently above or below it. Newton similarly maintained that, while bodies on earth appear only to descend under the attractive force of gravity, the fact of the matter is that those bodies are attracted to the center of a larger body and, therefore, could just as easily be conceived of as ascending. A third feature is the imputation of uniformity to all market forces. Newton similarly maintained that both the terrestrial and the celestial realms are subject to the same universal force.

Other scholars have also emphasized that Smith transferred the ubiquity of gravity, as the universal glue, to sympathy, which springs up naturally, most obviously in parental love. A mother feels pangs with the moans of her child. Sympathy is, as Lord Kames phrased it, the "great cement of human society" (quoted in Forget 2003, 284). Even the most hardened criminal is not exempt from such sentiments, suggesting again that sympathy is physiological and not a function of one's stock of virtue (see Smith 1790/1976, 7–13). An even more direct transfer of Newtonian imagery into Smith's moral philosophy can be found in the treatment of benevolence. Humans always have a capacity for benevolence, but, like gravitational attraction, that capacity diminishes with distance. Similarly, benevolence is strongest within the family and weak to nonexistent in the marketplace, where relations are mostly anonymous (see Otteson 2002, 298).[16] Other scholars, such as Norriss Hetherington (1983) and Deborah Redman (1997), have explored the Newtonian thrust in Smith's pronouncements on methodology, while Andrew Skinner (1967) and Paul Wood (1989) have traced Smith's appreciation for natural history.

I will here extend this line of reasoning into other facets of Smith's economic thinking. The first is his use of concepts from experimental physics and physiology, such as his treatment of labor and sympathy in terms of subtle fluids. The second is his appreciation of natural history and its ramifications for his appeals to market adjustments and the broad sweep of time. The third is his assimilation of the Copernican system in his distinction between appearance and reality in the moral realm. I develop these three themes here because they are relatively original, but by no means will I exhaust the ways in which Smith debited nature's bank.

We have already noted that Stephen Hales's *Vegetable Staticks* (1727) established that air could be "fixed" in assorted substances, such as amber, beeswax, blood, and water.[17] This in turn drew attention to the fact that air might have different properties, a recognition that enabled Joseph Black, in 1754, to make the fundamental discovery of "fixed air" in his dissertation on magnesia alba. Commentators have often noted the two senses in which Smith advances a labor theory of value. He speaks in some places of the value of goods being determined by what labor can command and in others of the value of goods being determined by the amount of labor embodied in them. It is the latter sense that was more enduring, particularly in the works of David Ricardo and Karl Marx. But what is perhaps more interesting is that Smith depicted the stuff of labor (imagine a subtle or imponderable fluid) implanting itself in objects and being stored there for later extraction. As he notes in *The Wealth of Nations:* "The labour of the manufacturer *fixes* and realizes itself in some particular subject or vendible commodity, which lasts for some time at least after that labour is past" (emphasis added). This image serves Smith well in drawing a distinction between productive and unproductive labor: "The labour of the menial servant, on the contrary, does not *fix* or realize itself in any particular subject or vendible commodity. His services generally perish in the very instant of their performance, and seldom leave any trace or value behind them, for which an equal quantity of service could afterwards be procured" (Smith 1776/1976, 1:330; emphasis added). In sum, any labor that is ephemeral, that cannot fix itself in an object for a later use, is unproductive. This encompasses "the declamation of the actor, the harangue of the orator, or the tune of the musician, [because] the work of all of them perishes in the very instant of its production" (331).

This image of fluid-like substances fixing themselves in grosser objects would have been a process familiar to Smith firsthand from his interchanges with Cullen and Black and secondhand from his reading of Hales. Moreover, only certain bodies could permit the fixation of air, just as only certain working contexts could be productive. Clearly, Smith thinks of labor as a substance, insofar as it can be transferred, stored, and extracted—and evaporate, for that matter. While we have no record of the genesis of this, his new approach to labor (there is no trace of it in his earlier lectures at Glasgow), it seems reasonable to assume that Smith utilized the conceptual tools of contemporary chemists. That the subtle substances of heat and air could be stored in gross bodies and then retrieved at a later time has an important analogue in the way in which labor was conceived to be like a vital fluid, which could be stored in objects and then extracted at a later point in time if it had not "evaporated."[18]

Smith's treatment of labor is strikingly original and not to be found in Locke and Hume, although both broach the idea that labor creates value. For Locke, labor can be "mixed" or "annexed" to other objects, especially land, but he thinks of this process primarily in terms of the creation of property rights (Locke 1764/1980, 19). That is to say, we have property in our own persons, actualized by our labor, and this can be transferred to land to create estate.[19] Hume challenges this position, noting in a footnote in book 3 of the *Treatise:* "We cannot be said to join our labour to any thing but in a figurative sense. Properly speaking, we only make an alteration on it by our labour" (Hume 1739–40/2000, 324n). The Physiocrats, likewise, believed that labor can only alter, and not create, goods and that it certainly does not embed itself in objects or create value. Smith, it appears, was the first to view labor as an alienable stuff that could, in the case of productive labor, be packed into other objects.

This might all sound somewhat strange if we did not have good reason to believe that Smith also thought of human sympathy in terms of a vital fluid. As Chris Lawrence has suggested, Edinburgh physicians such as Alexander Monro and John Gregory developed in the early 1750s "a model of the body in which sensibility, a property of the nervous system, predominated. . . . This they did by using the notion of 'sympathy'—which was no more than the communication of feeling between different bodily organs" (Lawrence 1979, 27). But even more fascinating is the effort to conceptualize the nervous fluid as an electric one.[20] This conceptualization was used to explain why feelings were communicated, not just within a body, but between different persons. As Gregory observed: "People subject to hysteric fits will often by sympathy fall into a fit by seeing another person fall into the same" (quoted in Lawrence 1979, 28).

Lawrence links this model of the body to Adam Smith's work on sympathy, emphasizing that these communications of the nervous system transpire physiologically (see Lawrence 1979, 28–31). Smith underscores, not only the rapidity with which sympathy springs into being, but also the physical reactions that it engenders. The sight of someone undergoing torture leads us to "tremble and shudder"; the sight of someone about to be whipped makes us "naturally shrink and draw back our own leg or our own arm" (Smith 1790/1976, 9–10). Because our eyes are our most delicate organ, the sight of someone with sore eyes induces in us a similar soreness (10). There is, as numerous scholars have emphasized, a physiological element to Smith's conception of sympathy (see Rothschild 2001; and Forget 2003). The fact that he might have conceived of labor as a transferable bodily fluid clearly fits within this broader physiological framework.

Paul Wood (1989, 89) has argued that one of the most prominent pursuits of the Scottish Enlightenment was natural history. Smith was instrumental in importing the ideas of René-Antoine Réaumur and Buffon, noting in the 1756 letter to the *Edinburgh Review:* "None of the sciences indeed seem to be cultivated in France with more eagerness than natural history" (Smith 1795/1980, 248–49).[21] This and what we know about the contents of Smith's library point to a keen and sustained interest in the subject. We know that he much admired Linnaeus and that he had read the *Systema natura,* for he makes use of it in "Of the External Senses" (most likely written before 1752).[22] He also had in his library Benjamin Stillingfleet's 1759 translation of Linnaeus's *Oeconomia naturae* and may well have read the original 1749 Latin version. Smith also notes the adaptiveness of species and makes an oblique reference to classification by genera and species in "The History of Astronomy." This supports the view that he had perused the *Oeconomia naturae* in his formative years (Smith 1795/1980, 38).

In *The Theory of Moral Sentiments,* Smith inserts a lengthy passage on the oeconomy of nature that recognizes that "self-preservation and the propagation of the species, are the great ends which Nature seems to have proposed in the formation of all animals" (Smith 1790/1976, 77). We are bestowed with various instincts that motivate actions with unintended consequences: "Hunger, thirst, the passion which unites the two sexes, the love of pleasure, and the dread of pain, prompt us to apply those means for their own sakes, and without any consideration of their tendency to those beneficent ends which the great Director of nature intended to produce by them" (78). It is language and our propensity to "truck, barter, and exchange" that sets us apart from all other species and gives rise to commerce and trade (Smith 1776/1976, 1:30). But, in every other respect, we are seamlessly joined to the richer oeconomy of nature.

Linnaeus had noted "the wonderful disposition of the Creator, in assigning to each species certain kinds of food, and in putting limits to their appetites" (Linnaeus 1791/1977a, 95). Smith possibly echoes these sentiments in his own ascription of a circumscribed stomach: "The desire of food is limited in every man by the narrow capacity of the human stomach" (Smith 1776/1976, 1:181). But much more may have been drawn from the Linnaean concept of an oeconomy of nature. Presumably, Linneaus's economic processes, such as the geometric rate of reproduction, the supply of and demand for food, and the mechanisms by which equilibria are restored, were not lost on Smith (see Schabas 1990a; and Müller-Wille 2003). For example, in Smith's analysis of the adjustment process that is unleashed by the sudden increase in the demand for black cloth generated by a public mourning (Smith 1776/1976, 1:76–77), we see many of the same supply-and-demand adjustments that Linnaeus identified in the case of a sudden fluctuation in the population of a given species. Resources are shifted in response to the disturbance, and equilibrium is once again restored. To be sure, equilibrating processes were already apparent in earlier economic tracts, but it is only with Smith that demand is given explicit treatment (see Hollander 1973, chaps. 1, 4).

Smith's 1756 letter to the *Edinburgh Review* also comments on the early volumes of Buffon's monumental *Histoire naturelle,* "one of the most widely read scientific works of the eighteenth century" (Wood 1989, 89). Going beyond the sheer fascination with natural history common at the time, Buffon proposed nascent theories of biological evolution and ascribed substantial antiquity to the earth. Needless to say, this caused quite a stir, both for the scientific implications and for the theological. Smith's letter was noncommittal. Buffon's system, he wrote, "is almost entirely hypothetical," although "supported or connected . . . with many singular and curious observations and experiments" (Smith 1795/1980, 248). While we do not know whether Smith was actually convinced by Buffon, as we have seen in "The History of Astronomy," his methodological predilections placed considerable weight on connections between disparate observations. Certainly, Smith, and for that matter Hume, sought to distance himself from the French intellectuals because of their alleged atheism, so it is not unlikely that his noncommittal stance on Buffon was similarly motivated. Clearly, he felt compelled to announce the work to his fellow Scots, and he helped inspire others, such as Adam Ferguson and Lord Kames, to cultivate a natural history of humankind in the tradition of Linnaeus and Buffon (Skinner 1967; Wood 1989; Koerner 1999), a project that entailed tracing the roots of human nature back to earlier stages of development. Smith's letter also acknowledged Rousseau's appeals to the "noble savage" (Smith 1795/1980, 251), and there are a number of similar appeals in *The Theory of Moral Sentiments*, for example, to the dignity of African slaves or the courage of aboriginal Americans (Smith 1790/1976, 205–9).

There is also a strong predilection in Smith, as in Buffon and Hutton, toward viewing the economy developmentally and thinking in large chunks of time. Such an approach was relatively novel in the history of economic thought—not the reference to historical events per se, but the effort to think of the flow of wealth from nation to nation as proceeding over a period of one or two, if not several, centuries. As we saw in the preceding chapter, Hume had already begun to think expansively in terms of time and, at least in his correspondence, had observed that no country could sustain its economic hegemony forever (Rotwein 1970, 200). Smith injects a more structured process into his economic history, one that goes far beyond the four-stages theory that captured his imagination and that of his contemporaries (see Meek 1971). Book 3 of *The Wealth of Nations* offers a lengthy analysis of "the natural progress of opulence," whereby countries or regions pass through several prolonged stages of economic development. While the evolution is still punctuated, there is a much more explicit sense in which the latter stages develop out of the former. Smith's conception is more epigenetic than preformationist. Also quite remarkably, Smith anticipated the decline of the British economy. He seems uncannily prescient, and quite unfettered by the colonizing temperament of his age, when, in the last sentence of *The Wealth of Nations,* he proposes that Great Britain must "endeavour to accommodate her future views and designs to the real mediocrity of her circumstances" (Smith 1776/1976, 2:947). Much of this seems in step with the efforts of Buffon and Hutton to extend the age of the earth and of life on it. One is reminded of the detachment that Darwin was able to muster in the closing sentence of *The Origin of Species,* transcending any anthropocentrism.[23]

Locke's *Second Treatise* had deemed the cultivation of the land to be man's mission in the state of nature; agriculture's historical priority made farming natural in a way that manufacturing was not. But an even deeper justification stemmed from the Physiocratic notion of the benevolence of nature. The mid-eighteenth century was a period when the emerging industrial order forced comparisons with agriculture as the locus of nature's gifts. Moreover, when it came to sorting out the paramount question of why some countries are rich and others poor, Smith peeled away the layer of disturbances wrought by human institutions and pointed to the underlying set of natural resources that ultimately determine what human labor, suitably divided, could unleash.

From Hutton, Smith would have garnered a greater appreciation of the earth's fertility and become acquainted firsthand with experimental agriculture. According to Simon Schaffer, Hutton spearheaded the development of rational agriculture in eighteenth-century Britain. Several of Smith's patrons and peers, including Henry Home, Lord Kames, and Cullen, were also part of this movement to foster the systematic study of agriculture (see Schaffer 1997, 138–41). Hutton had deemed agriculture the most virtuous of activities insofar as it gave humans, like God, dominion over their natural kingdom. It does not seem that Smith was so persuaded since mercantile trades tended to breed better the virtues he ranked most highly, namely, honesty, probity, prudence, and self-command (Smith 1790/1976, 63, 146; see also Rosenberg 1990). But even retiring merchants tended to join the landed gentry, in part because of the security that comes from keeping their wealth ever in view (Smith 1776/1976, 1:377–78). Insofar as the primary motive of all human activity is to seek the approval of others, this also tips the balance in favor of the agrarian sector since its wealth was easily displayed.

From a broader perspective, Smith deemed agriculture more valuable: "Agriculture is of all other arts the most beneficent to society, and whatever tends to retard its improvement is extremely prejudicial to the public interest" (Smith 1978, 522). This was partly due to the natural progress of opulence, whereby commerce and urbanization follow the cultivation of the nearby land (Smith 1776/1976, 1:377). The agricultural sector was the most productive, per unit of capital invested, because of the added assistance of nature: "No equal capital puts into motion a greater quantity of productive labour than that of the farmer" (Smith 1776/1976, 1:363; see also McNally 1988). Smith even attempts to quantify nature's contribution, a calculation subsequently undertaken, as we will see in chapters 6 and 7, by Ricardo and John Stuart Mill: "It is the work of nature which remains after deducting or compensating every thing which can be regarded as the work of man. It is seldom less than a fourth, and frequently more than a third of the whole produce. No equal quantity of productive labour employed in manufactures can ever occasion so great a reproduction. In them nature does nothing; man does all" (364). The gift of nature could not be more sharply contrasted with the feeble efforts of human manufacturing. In sum, Smith's Physiocracy, at least in this respect, drew considerable support from Enlightenment natural history.

J. Ronnie Davis (1990) and Charles Griswold (1999), among others, have emphasized the significance of deception in Smith's moral philosophy. First and foremost, there is self-deception, "the source of half the disorders of human life" (Smith 1790/1976, 158). But deception also operates at the macroscopic level. The invisible hand is also a sleight of hand. The lower orders toil because they believe, falsely, that riches are worth the effort when, in fact, they are apt to lose what serenity they do possess: "It is this deception which rouses and keeps in continual motion the industry of mankind" (Smith 1790/1976, 183). At the individual level, then, there is no guarantee of greater fulfillment. In the aggregate, however, society definitely tends toward greater happiness, not because industry begets more trinkets, which satisfy only in a hollow sense, but because it begets greater liberty and virtue (Griswold 1999, 262–66; Rothschild 2001). This in itself strongly suggests that, for Smith, the moral world was designed to bring about the most benevolence. And, of course, it also had built into it all sorts of tendencies toward greater general prosperity over time, through the division of labor and trade.

Interestingly, the first acknowledgment of the invisible hand is in "The History of Astronomy," and there the hand is that of Jupiter (see Smith 1795/1980, 49; Ahmad 1990). Smith would have gathered from the triumph of the Copernican system over the Ptolemaic that the heavens themselves are so construed as to be one grand deception. After all, the sun does not really rise and set every day, whatever the common people might think. The new astronomy implied that the world was designed in a more complex fashion, thereby creating certain illusions of self-centeredness (the earth taken to be the center of the universe) and certain illusions of stability (the earth taken to be at rest). All this had been threshed out among natural philosophers of the seventeenth century. Kepler and Newton provided the most satisfactory account of the heliocentric system, while Galileo and Descartes, with their distinction between primary and secondary qualities, established the immense gulf between the physical world and the internal world of the mind. The given of experience—sights, sounds, and the like—was produced by pure matter, the collisions of microscopic particles that, because they were devoid of any perceptible properties, could be understood only mathematically.

George Berkeley attempted to restore a commonsense metaphysics by proclaiming that only that which is perceived can exist. His famous motto, *Esse est percipii* (To be is to be perceived), assumed that the material substratum, the swirl of atoms, simply did not exist.[24] Hume in certain respects shared Berkeley's aversion to the mechanical philosophy and also sought to restore a reality that matches that of our experience. But his denial of the existence of the material substratum was less emphatic. He was not an immaterialist like Berkeley, merely a skeptic about how much we could discern about the microscopic world. And he did not share Berkeley's wish to reduce metaphysics to the view of a common gardener (see Berkeley 1734/1979, 67–69). Quite the contrary, he deplored "the vulgar, who take things according to their first appearance." Philosophers, on the other hand, recognize that "in every part of nature there is contain'd a vast variety of springs and principles, which are hid, by reason of their minuteness or remoteness" (Hume 1739–40/2000, 90).

When Smith first read Descartes, Berkeley, and Hume, whether at Oxford or after his return to Scotland, remains a matter for conjecture, but their influence on his ideas is unmistakable, not least because he cites them explicitly. It seems difficult to believe, then, that Smith's rich insight that the moral world is full of deception owes nothing to these philosophers and their own struggles with the implications of the heliocentric theory. True, the trope of deception was also prevalent in Mandeville's *Fable of the Bees,* but it was only skin deep, so to speak.[25] Smith's appreciation of the divide between appearance and reality is at the core of his metaphysics. This position is bolstered by the contrast between Mandeville's play of vice and Smith's much more benevolent world, whereby the virtues of industry and prudence might induce a more harmonious, if not happier, world.

Even more profound was Smith's realization that all the "bustle of the world" was for nought. In *The Theory of Moral Sentiments,* there is a passage depicting a man of little means whom "heaven in its anger had visited with ambition." The man works incessantly to move up the ladder of social rank and income, to endear himself to "those whom he hates," and to be "obsequious to those whom he despises," only to discover "in the last dregs of life, his body wasted with toil and diseases, his mind galled and ruffled by the memory of a thousand injuries and disappointments which he imagines he has met with from the injustice of his enemies, or from the perfidy and ingratitude of his friends, that he begins at last to find that wealth and greatness are mere trinkets of frivolous utility" (Smith 1790/1976, 181). What really matters, "ease of body, or tranquillity of mind," is as elusive as wisdom. Berkeley had made a concerted effort to demonstrate that God could not be a deceiver. For Smith, he could not be otherwise. The centrality of the appearance/reality distinction in his writings owes much to the broader metaphysical debates of the early modern period.

SMITH AND RELIGION

Smith's religious commitments are, perhaps, more disputed than any other facet of his thought.[26] As Emma Rothschild has emphasized, this is partly due to his idiosyncratic use of ancient (mostly Roman) and Christian elements (Rothschild 2001, chap. 5), which renders murky any attributions of a providential order. While most opt for Smith as a lukewarm theist, there are some, such as Peter Minowitz (1993), who deem him to have been an atheist, at least in his later years, and specifically in those years when he was composing *The Wealth of Nations*. Minowitz concedes that the earlier *Theory of Moral Sentiments* is favorably disposed toward the operations of a deity, but he also points to what he sees as a strong anticlerical tenor throughout the work.[27] One might grant the anticlericalism, but it would be hard to make sense of the frequent references to the intentions of "Nature" or the less frequent references to a god, either Christian or Stoic, if Smith had been an atheist. Ultimately, however, it was not imperative that a treatise on political economy address theology, so the absence of specific reference to the subject in *The Wealth of Nations* tells us little.

There are some at the other end of the spectrum, James Otteson most recently, who view Smith as a full-bodied Christian. According to Otteson, Smith believed that the most virtuous persons were Christians (see Otteson 2002, 255–57). But this squares with neither Smith's moral theory, which, apart from an endorsement of the Golden Rule, steers away from Christian doctrine, nor his persistent deprecation of Europeans for their deplorable practice of slavery, let alone their tendency toward foppery and cowardice: "The frivolous accomplishments of that impertinent and foolish thing called a man of fashion, are commonly more admired than the solid and masculine virtues of a warrior, a statesman, a philosopher, or a legislator" (Smith 1790/ 1976, 63; see also 206). Smith is unique among British philosophers in his praise of aboriginals and Africans, particularly of their courage when tortured or shackled (see Griswold 1999, 198). Bravery in the face of death ranks high in Smith's moral firmament. And, if there was a single European Smith most admired for his equanimity while dying, it was *le bon David,* who, after all, refused to receive Christian rites. How could Smith praise Hume's death, let alone those of pagans, and still be a devout Christian?

Others such as John Dunn view Smith's close friendship with Hume as a compelling reason to put them both in the same category, that of the "practical atheist" (Dunn 1983, 119). But their respective trajectories, both lived and written, are sufficiently dissimilar to suggest that ascribing identical religious beliefs to them is likely to be misleading. Smith puts considerable weight on the presence of a planned world, whereas the gist of Hume's posthumous work is that we must renounce such beliefs as fallacious. Even if Hume had not left behind his *Dialogues concerning Natural Religion,* his essay "Of Suicide" would still point to a significant break with Christian views. Smith's own comments on the topic of suicide, added to the sixth edition of *The Theory of Moral Sentiments* (1790), show that, unlike Hume, he was not willing to see it as a natural occurrence (Smith 1790/1976, 287). Moreover, his reluctance to support Hume on two critical occasions suggests that, if he was an atheist, his beliefs were not strongly motivated. Not only did he fail to endorse Hume's candidacy for a post at the University of Glasgow precisely because of Hume's alleged atheism, but he also did not honor a long-standing obligation to serve as Hume's literary executor, which entailed shepherding the *Dialogues* into print (see Ross 1995, chap. 17). Of course, that Smith reneged on his promise could imply only that he was highly strategic. Still, if he had embraced atheism, he might well have been more zealous about taking a stand, as Hume had done.

It seems more prudent to view Smith as a deist and Hume as an agnostic. While both used conventional language about religion, and neither ever denied the existence of God altogether, Smith makes many more references to the deity or to Providence than does Hume, which could mean that his position was closer to that of a deist like Voltaire (see Haakonssen 1998, xv). There is a strong sense throughout Smith's works that the world was designed by a deity or "Great Author." But even more telling, perhaps, is the allusion to purpose or a telos, which would seem to make Smith a theist. Lisa Hill has argued that, in Smith's metaphor of the world as a watch, "God is cast in the role of the watchmaker, the First Cause and architect of the watch's Final Cause or purpose" (Hill 2001, 10; see Smith 1790/1976, 87). Hill discerns a Stoic theodicy in Smith, albeit one that has acquired a utilitarian gloss. Happiness and material prosperity are the ends intended by our planner, for humans are certainly "God's crowning achievement" (Hill 2001, 13).

There is much merit in this view, but it downplays the strongly pessimistic component of Smith's scheme, the sense in which we are but pawns on a chessboard, subject to vanity, greed, and ambition. If, at bottom, we are driven by nothing more than the desire to garner the approval of others, doing so primarily by means of displays of wealth and fashionable trinkets, as Smith argues so forcefully in the first book of *The Theory of Moral Sentiments,* then society is rendered a mere hall of mirrors. Smith's caustic observations on Louis XIV, who was considered universally to be "the most perfect model of a great prince," are telling. Such admiration was won, not by "extensive knowledge, by his exquisite judgment, or by his heroic valour," qualities altogether lacking in the French king, but by countenance and mannerisms alone. Such behavior as Louis's found among anyone of a lower rank would have been considered ridiculous, downgrading as it did the higher virtues: "Knowledge, industry, valour, and beneficence, trembled, were abashed, and lost all dignity" (Smith 1790/1976, 54). In more than one passage, Smith deplores the fact that fashion and fame have eclipsed the nobler pursuits of virtue and wisdom.

Much of the debate over Smith's religious views centers around his famous metaphor of the invisible hand. I side with Rothschild's argument "that the invisible hand was an unimportant constituent of Smith's thought" (Rothschild 2001, 136). The main problem among Smith scholars is explaining the metaphor's posthumous fame, not its function in Smith's thought (besides its "introduction" in the posthumous "History of Astronomy," it appears only once in *The Theory of Moral Sentiments,* only once in *The Wealth of Nations,* and never in the correspondence). The very fact that Smith first connects it to the caprice of Jupiter in "The History of Astronomy" suggests that he was not striving for a consistent or well-thought-through metaphor. Oddly, most scholars also downplay its appearance in *The Theory of Moral Sentiments,* where it generally serves to justify the inequality of wealth, and emphasize its use in *The Wealth of Nations* to justify the existence of a spontaneous order. More to the point, if the invisible hand is meant to represent divine action, then Smith would have had in mind a Christian God. Yet the evidence still points more toward a deity with strong Stoic vestiges, and such a deity would be very removed from the human stage—and certainly lack a hand, let alone a body.

If the invisible hand has significance, that significance is, as Rothschild (2001, 145) avers, its policy implications. It is far better to allow nature to take its course and trust the social order to individual pursuits and judgments than to turn it over to government planners. Yet even this popular rendition requires considerable qualification, insofar as Smith recognizes the role of governments in providing public goods and, more paternalistically, guarding against the deterioration of the minds and bodies of their subjects (Smith is gravely concerned about the decline of the martial spirit in his time). But this popular rendition becomes even more problematic once one reflects on how cynical Smith is about the human condition and man's inability either to know himself or to form good judgments. As we have seen in the epigraph to this chapter, Joseph Cropsey makes the point that, as with Hume, so too with Smith passions outweigh any calculations of the mind (see Cropsey 1975, 143).

Individual deliberation plays a decidedly subordinate role in Smith's

Wealth of Nations. A. L. Macfie has suggested that Smith's invisible hand demonstrated that the good of society outweighed that of the individual: "The belief, so popularly accepted in the economic world, that Smith was primarily an individualist, is the very reverse of the truth. For him as for Hume, the interests of society were the end. By all means let the individual be encouraged to chase 'trinkets,' so long as this conduced to that end" (Macfie 1961, 23). Dobb (1973) made a somewhat different point when he emphasized the methodological holism of the classical economists, including Smith; economic classes were taken to be the unit of analysis, not, as in the neoclassical period, economic agents. Smith tends always to talk of groups such as merchants or bankers. More recently, Andy Denis has raised the question about Smith's allegiance to individualism and argued quite persuasively that "Smith systematically denies the autonomy of the individual with respect to the whole of which he or she is part. For Smith, individual liberty is not the end, but the means, of sustaining social order and property" (Denis 1999, 71).

This does not undercut the individual pursuit of virtue, of self-command most saliently. The impartial spectator that lies within every bosom forces us to reflect and, one hopes, opt for the more virtuous path. But it tends to do its work post facto, at the end of the day, after our deeds are done. By and large, we are driven by more fundamental passions: "When we are about to act, the eagerness of passion will seldom allow us to consider what we are doing, with the candour of an indifferent person" (Smith 1790/1976, 157). It is only when we ascend to a state of reflective equilibrium, to speak anachronistically, that we begin to act correctly. Moreover, the impartial spectator steers each of us toward a universal set of moral rules and norms. Smith makes for a good proto-Kantian. The wise and virtuous person acts with great "constancy and firmness," merging altogether with the impartial spectator, who knows right from wrong (146–47). Yet this same person with good Stoicism also grasps his true insignificance, as but a speck in the universe, "as an atom, a particle, of an immense and infinite system" (276).

Even in Smith's political economy the individual is diminished. The phenomena that unfold in the economic realm—prices, trade, and commerce—do so not because of specific individual differences or deliberations, as might be maintained in neoclassical economics. Rather, economic development is the result of deeper forces overriding that which makes us individuals. As Denis has observed: "Smith's individualism extends so far as the individual is a representative of property, and no further" (Denis 1999, 80). Each of us instantiates a history of property relations, but, in many ways, simply as a way station on the road to the production of yet more wealth. Even the engine of economic growth, the ongoing division of labor, results, not from design, or from intention, but from the propensity to exchange. Economic growth is, first and foremost, a function of the size of the market, and markets grow with the natural progress of agriculture. Similarly, the "desire of bettering our condition" that also motivates the accumulation of wealth (Smith 1776/1976, 1:341) is not something we choose or acquire. It is there from "cradle to grave" and becomes effectively tautological.

CONCLUSION

As we have seen, the view that Smith's political economy has a natural foundation is well established. For Smith, economic phenomena were closely wedded to physical and biological nature, and his construal of these phenomena reflects their links to nature. Whether in his treatment of sympathy and labor or in his analysis of price mechanisms, Smith drew on recent developments in the natural sciences. Furthermore, human nature was more governed by instinct and animal-like passions than by reason. Smith also retained the Physiocratic predilection for agriculture over manufacturing, suggesting that, qua economic creatures, humans were first tied to the land. Smith also treated economic phenomena as at one with developments in the natural world. There are clear signs of the same temporal span as found in Buffon or Hutton. Finally, the world that we inhabit is replete with illusion and deception, and this can make sense only given the much more anchored commitment to the divorce between appearance and reality that came to the fore after Copernicus. This is hardly surprising, given Smith's deeply rooted belief that the world is designed and providential (see Viner 1991; Hill 2001). Our human oeconomy would necessarily sit in harmony with the oeconomy of nature. The invisible hand is not so hidden.

CHAPTER SIX

Classical Political Economy in Its Heyday

Ricardo conquered England as completely as the Holy Inquisition conquered Spain. Not only was his theory accepted by the city, by statesmen and by the academic world. But controversy ceased.

—John Maynard Keynes, *The General Theory of Employment, Interest and Money*

Smith's *Wealth of Nations* cast such a formidable shadow over the study of political economy that no comparable text in the English language appeared until David Ricardo's *Principles of Political Economy and Taxation* (1817). But Ricardo's shadow might have stretched even farther. Numerous scholars have underscored the radical recasting of the subject by Ricardo (1772–1823) that dominated the field for several decades.[1] More than any other work either in its day or, arguably, in the history of the discipline, Ricardo's *Principles* set the substantive and methodological parameters of the discipline that we know today as economics.

Ricardo centered his analysis on the question of the distribution of wealth among the three classes—landowners, capitalists, and laborers. While he also addressed policy matters, particularly taxation, his primary objective was to set down the universal laws that governed the production and distribution of wealth. As Halévy observed, his emphasis on the operation of laws constituted one of his most significant innovations (Halévy 1972, 267). Smith had noted tendencies and certainly sought out general principles, but he rarely spoke of laws per se. Perhaps more startling was the fact that Ricardo's ascription of laws in the economic realm went unjustified. In other words, he took their existence as self-evident.

In this he was much indebted to Smith, who had enriched the scientific standing of political economy and engendered a widespread belief in an economic order. Political economy was also beginning to make its way into elite scientific circles, a development that Ricardo's own efforts greatly fueled. By the 1830s, the science of political economy was blessed by the British Association for the Advancement of Science and the major encyclopedias; indeed, articles on political economy were often placed among the sections on the "hard sciences." Political economy had also insinuated itself into the lectures and texts of the leading scientists of the age, most notably John Herschel, William Whewell, and Charles Babbage.[2] There were certainly critics of political economy, but they were mostly of the Romantic persuasion, such as Samuel Taylor Coleridge and Thomas Carlyle, and, thus, critical of Enlightenment science at large. And those who attacked Ricardo for his overly deductive approach—Richard Jones most notably—were still keen to enhance the scientific standing of political economy (see Collini, Winch, and Burrow 1983, 80).

Political economy shared much the same allure as geology at the time. It was not accidental that Ricardo and Richard Jones both belonged to the Geological Society or that George Poulett Scrope drew direct analogies between the two fields in his theoretical work. Indeed, Charles Lyell forged close friendships with both Scrope and Nassau Senior and for one year attended J. R. McCulloch's lectures on political economy (see Rudwick 1974, 1979; and Rashid 1981b). Both political economy and geology were seen as forces for secularization, the one enhancing commercial ends, the other challenging scriptural authority on the age and creation of the earth (see Hilton 1988, 149). Both subjects were also embraced by Christian conservatives such as William Buckland since they gave a scientific foundation to Britain's claims to being the richest and most powerful European nation. Political economy justified the triumph of Britain's commercial order and manufacturing sector, while geologic inquiries reinforced the fact that Britain was geographically blessed. Her mineral resources and rich array of internal waterways more or less guaranteed prosperity.[3]

Ricardo's *Principles* skirted the broader questions of human motivation and moral agency and presented the theory of value and distribution in a manner akin to that in which Euclid's geometry was presented. Deduction rather than induction ruled the day. Indeed, Ricardo often remarked in his correspondence that political economy was a science like mathematics, not like morals.

He never used more than simple algebra and hypothetical numerical examples to illustrate his main principles, but his readers always had the sense, as Alfred Marshall was later to remark, that he was groping his way toward the calculus. Moreover, in one limited sense, his theory of rent, as Mark Blaug has pointed out, embodied the concept of marginal analysis (see Blaug 1972, 276).[4]

Despite his predilection for rigorous analysis, Ricardo was not indifferent to the political and ethical dimensions of economic analysis. As several authors have shown, he had cultivated his own distinct cast on these topics, initially under the influence of the radical philosophers Jeremy Bentham and James Mill.[5] Moreover, as a member of Parliament from 1819 to 1823, he had the opportunity to speak to these issues firsthand, as he had previously at meetings of the Political Economy Club and in the pages of the *Edinburgh Review*. Ricardo was a more well-rounded individual than his *Principles* taken on its own would suggest. Nevertheless, his treatise, insofar as the subject is presented in an abstract, deductive, and ahistorical fashion, set the format for mainstream theory right up to the present time. If Smith was the Newton of political economy—for Newton's work was also full of asides, half-baked analyses, and mysterious segues—Ricardo was its Euler. Analytic rigor, clarity, and consistency had become the desiderata.

Ricardo is celebrated now, as he was in his own day, for weaving together an integrated and logical theory of value and distribution. Even so, in that theory one still finds vestiges of the eighteenth-century emphasis on physical nature. Alas, hardly any scholars have asked how Ricardo conceived of an economy or even if he perceived such a construct. The debate has generally been confined to the internal textual logic of the period from Smith to Ricardo, predicated on the assumption that Smith and Ricardo had identical conceptions of the economic order. The vast secondary literature on the economics of David Ricardo focuses almost exclusively on such issues as whether he had a consistent and determinant theory of value and prices, whether he positioned the tendency for the rate of profit to diminish over time at the center or at the periphery of his analysis, and whether he had a corn model implicit throughout his mature works. There are even several articles and a book devoted simply to the task of evaluating the conflicting interpretations of Ricardo.[6]

Keith Tribe and Maxine Berg are two exceptions. They have addressed the broader issues that concern me here, although with views that I do not entirely endorse. Tribe suggests that Malthus had closer ties to eighteenth-century naturalism than did Ricardo (Tribe 1978, 127) and, indeed, sees little in Ricardo that still resembles that earlier train of thought. I will suggest here that there are still some remnants of appeals to physical nature, but certainly fewer than are found in Malthus. Berg describes Ricardo as "a missionary in a land of pagan naturalists," by which she means that he alone envisioned the unlimited potential of mechanization (Berg 1980, 65). This claim I find overblown. Berg also maintains that Ricardo "drew a sharp distinction between this 'natural world' and the socio-economic world he was attempting to analyse" (47). Again, I would qualify that and submit that the sharp distinction came only with John Stuart Mill. Indeed, Berg also acknowledges Ricardo's frequent appeals to the laws of nature and suggests that, for Ricardo, an unregulated economy "would gravitate to the operations of the 'laws of nature'" (46). But she claims these allusions served a "negative purpose" insofar as they detracted from the image of technical progress (47). This tension in Ricardo will be explored later on in this chapter.

THOMAS ROBERT MALTHUS

Although commentators often bemoan the lengthy temporal gap that separated Smith from Ricardo and the resultant perceived delay in the development of classical economics, they underestimate the contributions of Malthus. His *Essay on the Principle of Population* (1798), while initially not much more than a pamphlet, nonetheless left its imprint on political economy, particularly the enlarged second edition of 1803. Although the principle of diminishing returns was not explicitly articulated until the fifth edition of 1817, it was implied from the very start. But the main contribution of the *Essay* was to bring attention to bear on the problem of scarcity, not only of land, but also of capital. Scarcity became, as many have argued, the backbone of classical thinking, especially insofar as it was conceived of in historical terms, in terms of the falling rate of profit (see Hollander 1987, 194–202; and Winch 1996, 270–72).

As the first professor of political economy in England, at the East India College at Haileybury (1804), and as a founding member and active participant in the Political Economy Club, Malthus had a sustained interest in economic debates and theory. The 1798 *Essay*, however, predates his specific cultivation of economics. It was intended for a general audience, as a challenge to William Godwin's claims regarding the perfectibility of mankind. Unfettered population growth was simply a means to demonstrate that unfortunate end. Indeed, Malthus knew full well that the key insights about population were already to be found, at least in passing, in the works of Montesquieu, Hume, and Smith. Robert Wallace had outlined some of the basic principles of Malthus's argument as early as 1753, and we know that Malthus had read Wallace. Historians have subsequently unearthed numerous additional predecessors, with the result that Malthus appears to have said nothing new.[7] In the history of science, however, timing is all. Moreover, in subsequent editions of the *Essay,* Malthus exercised considerable charity in acknowledging his predecessors and contemporary critics.[8] But he also greatly expanded his argument, both on the subject of positive and preventive checks to population growth and on that of the moral and economic dimensions of the problem.

Malthus predicates the entire argument of the *Essay* on two fundamental postulates: that food is necessary to human existence and that sexual passion will persist unabated. Given the potential for population to grow at a geometric rate and a most optimistic estimate of an arithmetic growth rate for agriculture, it follows that population will soon outstrip food supply. But Malthus believed that mankind was still far away from the point of saturation. Population had rarely, if ever, grown geometrically, and many other factors, such as war and disease, capped its expansion.

Interestingly, Malthus maintained that what would most accelerate our course toward the dire end of standing room only would be a greater demand on the part of workers for goods and services. Malthus suggests at one point that, were production stimulated by a full desire among laborers for "the necessaries and conveniences of life," then there might be no practical limit to the power of production and that population would have already reached a level tenfold what it was in 1820 (Malthus 1820/1989, 1:348). Implicit in this claim, then, is a fairly passive working class, one responding to signals rather than making independent decisions. It should come as no surprise, then, that Malthus tended to side with the more paternalistic policy measures of his time, thereby acquiring a reputation of being hard-hearted when it came to the lot of the poor.

The basic idea of a superfecundant population reaches back to Smith, if not to Petty. But only Malthus followed through on its implications when the scarcity of land was factored into the equation. The message of doom and gloom that resulted cast a dark cloud over political economy, the discipline being perceived as nothing but a bringer of bad tidings for decades thereafter. Carlyle's "dismal science" epithet gained currency in the 1830s, long after Malthus had issued his dire warnings (see Levy 2001). Indeed, the optimistic tenor found in Smith—who saw a modicum of happiness among the lower orders (at least in Western Europe)—was not truly revived until the 1870s.

My focus here is on Malthus's strong commitment to the uniformity of nature. First, he maintains that it would be impossible to develop a "human science" if nature were "fickle and inconstant." Indeed: "The constancy of the laws of nature, and of effects and causes, is the foundation of all human knowledge" (Malthus 1803/1989, 1:311). There are traces of Hume in these passages, although no explicit recognition. But, because Malthus embraced the idea of a providential order, his ascribed uniformity did not require an elaborate justification (2:88).[9] God's injunction by which we procreate and possibly encounter temptation as well as redemption was part of a greater plan. It was better, however, to focus on the "Book of Nature" than to begin to make sense of the intentions of the omnipotent mind of the deity. For Malthus, it would be arrogant "for any man to suppose that he has penetrated further into the laws of nature than the great Author of them intended, further than is consistent with the good of mankind" (2:227). Of course, this still gave rise to the question of how far God intended us to penetrate into nature or whether he intended us to know with relative certainty what is best for us, but Malthus did not pursue this line of inquiry.

Malthus also spoke of the "inevitable laws of human nature" (Malthus 1803/1989, 1:325). Our passions, he suggests, govern us in much of what we do with the same regularity as the laws of nature. It is foolhardy to believe that human institutions, such as those proposed by Godwin or Condorcet, could ever discipline us fully. Like nature itself, these laws must be obeyed: "Nature will not, nor cannot, be defeated in her purposes" (2:116). In the later *Principles of Political Economy* (1820), however, Malthus appears to have accepted greater scope for human agency. There, the laws of society, unlike the laws of physical nature, could be interfered with. The wise statesman, however, will be reluctant "to interrupt the natural direction of industry and capital" (Malthus 1820/1989, 1:20; see also 13). In that sense, then, Malthus still favored the idea of a natural course to economic development.

The *Essay* is also replete with analogies drawn from the world of plants and animals, presumably as a device supporting the overall argument that human fecundity is part of the natural order. In a passage that would later prompt Darwin to adumbrate the mechanism of natural selection, Malthus extends the confrontation between reproduction and land scarcity to the entire natural realm: "Through the animal and vegetable kingdoms Nature has scattered the seeds of life abroad with the most profuse and liberal hand; but has been comparatively sparing in the room and the nourishment necessary to rear them. . . . Necessity, that imperious all-pervading law of nature, restrains them within the prescribed bounds. The race of plants and the race of animals shrink under this great restrictive law; and the race of man cannot by any efforts of reason escape from it" (Malthus 1803/1989, 1:10). Insofar as we are all part of God's plan, we share much with those kingdoms, above all the principle of fecundity. In this sense, the principle of population is one of God's "great laws of nature" that must be obeyed (2:250).

Malthus views some of the positive checks to population—for example, drought, pestilence, and plague—as being due entirely to the laws of nature (Malthus 1803/1989, 1:19, 303). While we might well see all these as embodying human agency—as being the result, for example, of poor water conservation, excessive pesticide use, deforestation, or poor sanitary conditions—for Malthus they were purely due to natural forces, if not direct acts of God. Indeed, for Malthus, plague occurs because "we have offended against some of the laws of nature." We have allowed a "state of filth and torpor" to prevail to the point that God's intentions to bring happiness and virtue have been displaced. Fortunately, the plague cleanses us of those vices, inspires virtuous acts (the "removal of nuisances, the construction of drains, the widening of streets," etc.), and, thus, restores the health and happiness of the people (2:89). In that sense, offenses against nature are ameliorated by purely artificial means.

For Malthus, these inescapable laws of nature are essential to the telling of his population tale. He paints a picture whereby the natural predicament of a burgeoning population confronting a limited amount of arable land is unavoidable no matter how clever man might be. Even the climatic vicissitudes are helpful reminders of the strong forces of nature: "According to the natural order of things, years of scarcity must occasionally recur, in all landed nations. . . . The prosperity of any country may justly be considered as precarious, in which the funds for the maintenance of labour are liable to great and sudden fluctuations, from every unfavourable variation in the seasons" (Malthus 1803/1989, 1:385).

Agriculture alone sets the final limits to economic prosperity and, thus, to population. Trade notwithstanding, "the only true criterion of a real and permanent increase in the population of any country is the increase of the means of subsistence" (Malthus 1803/1989, 1:302). So taken is Malthus with the production of corn that he forgets even to include fish or fowl as sources of food. For him, as for the Physiocrats, wealth comes from the cultivation of the land: "Land, in an enlarged view of the subject, is incontrovertibly the sole source of all riches" (392). Nor should we forget our debt to nature's gift of fertile land: "If the earth had been so niggardly of her produce as to oblige all her inhabitants to labour for it, no manufacturers or idle persons could ever have existed. But her first intercourse with man was a voluntary present; not very large indeed, but sufficient as a fund for his subsistence" (393).

Even in a developed economy, the aggregate level of manufacturing is constrained by the surplus produce of the land "and cannot in the nature of things increase beyond it" (Malthus 1803/1989, 1:393). This surplus must also be advanced to the manufacturer before the work can be undertaken. Here, Malthus explicitly recognizes the views of the Physiocrats. He appears to side with them much more than with Adam Smith, who, perhaps unintentionally, granted manufacturing more significance than was warranted, or so Malthus maintains (405). In any event, agriculture is the bedrock of all healthy and lasting economies: "In the history of the world, the nations whose wealth has been derived principally from manufactures and commerce have been perfectly ephemeral beings, compared with those the basis of whose wealth has been agriculture. It is in the nature of things, that a state which subsists upon a revenue furnished by other countries must be infinitely more exposed to all the accidents of time and chance than one which produces its own" (395).

Malthus endorses Smith's proposition that an investment of capital in land is preferable to an investment in commerce or manufacturing, the former being more secure and permanent. He also maintains, contrary to Smith and Ricardo, that wages in the manufacturing sector are much less stable. It is, thus, better for a nation to build up its agrarian sector and, if it is also blessed with healthy commercial and manufacturing sectors, to strive for agrarian autarky (Malthus 1803/1989, 2:42). Malthus is, thus, at loggerheads with Ricardo, who advocated the extreme specialization of national economies in a world of unrestricted trade. Malthus recommends that a nation be self-sufficient in terms of food production so as not to be dependant on the whims of others.[10]

Commerce is also subordinate to agriculture. The former can flourish in the wake of the latter, but not vice versa (Malthus 1803/1989, 1:399). Indeed, even from "a commercial point of view," the most profitable venture is the sale of raw produce (405). Unfortunately, commerce has become distorted owing to monopolies and other "peculiar encouragements," and things have not followed their natural course. Commerce has increased beyond the surplus produce of the land, and, as a result, "the body politic is in an artificial, and in some degree, diseased state" (408). Malthus recommended swift action: reducing the import of luxury items such as tea, sugar, and coffee and returning to an agrarian economy based on the export of corn (409). He partakes somewhat of the Romanticism of the day, harking back as he does to the glorious time, from the late seventeenth century through the mid-eighteenth, when Britain had a flourishing agrarian economy.

This yearning for a bygone age is also evident in the discussion in the *Principles* of the different definitions of *wealth*. Malthus feels that such definitions have become far too broad and that it is important to return to one that views wealth as consisting of "those material objects which are necessary, useful, or agreeable to mankind" (Malthus 1820/1989, 1:28). He notes that not even Smith was sufficiently precise, insofar as he allowed some nonmaterial goods into the domain. Malthus himself opts for things tangible and observable. His construal of wealth has both Physiocratic and Smithian roots; the agrarian sector is the primary source of all riches.[11]

For all his bold and drastic warnings, Malthus regarded himself as a disciple of Smith. In his lectures at Haileybury, he stayed close to the outlines of *The Wealth of Nations* and initially intended his *Principles of Political Economy* to be a clarification of that book. Work on this project, which commenced in 1804, was interrupted by revisions of the *Essay* and other commissions. After engaging in many intensive debates on a wide range of topics with Ricardo (the two had met in 1811 [see below]), Malthus revised the text considerably so as to incorporate his refutations of Ricardo's doctrines. He struggled through several more drafts before releasing his manuscript to the press in 1820. The book did not sell very well; indeed, the publisher, John Murray, lost money and refused to issue a second edition. Malthus nevertheless undertook extensive revisions, although he never finished them to his satisfaction. A second edition, under the guidance of an anonymous editor, was published posthumously in 1836 (see Malthus 1820/1989, 1:lx–lxi).[12]

Some scholars, John Pullen most notably, have suggested that "it is more appropriate to situate [Malthus's] *Essay* within the general framework of [his] *Principles,* and to regard the *Essay* as a more intensive and extensive treatment of one particular aspect of the *Principles*" (Malthus 1820/1989, 1:xvii–xviii). That may well be the case post facto, but, historically, the influence of the *Essay* was orders of magnitude greater than that of the *Principles.* A more viable claim might be that Malthus considered himself, at least by the time of the second edition of the *Essay* (1803) and his appointment at Haileybury (1804), to have become a political economist and that, in that respect, his efforts at producing a major treatise on the subject, albeit much delayed, were integral to his mission.

But, without question, the principle source for Malthus's political economy was Adam Smith. The *Principles* was an explicit attempt to restore clarity to the Smithian doctrine by correcting the "confusions" wrought by Ricardo and Say. There is much evidence to suggest that Malthus was also favorably disposed toward the Physiocrats.[13] In that respect, the lineage between Smith and Ricardo is more pure, insofar as Ricardo steered away from French influences, with the possible exception of Say.

In 1811, Ricardo and Malthus became acquainted, and there began a close friendship. They met at regular family gatherings at Ricardo's country estate and carried on a regular correspondence that continued right up until Ricardo's death in 1823. Ricardo even left Malthus a sizable sum in his will. Yet the two differed, sometimes openly in print, on almost every point of economic theory and policy. Malthus never accepted Ricardo's version of the theory of rent, and they were persistently divided on the question of the Poor Laws, the Corn Laws, currency, and governance in general. Perhaps most puzzling of all is the fact that, given their persistent difficulties in persuading one another of their respective viewpoints, they remained on such good terms. There seems to have been an almost perverse element in their desire to confront each other on both theoretical and policy issues. Hollander has aptly characterized this as Malthus's "love-hate relationship" with Ricardo (Hollander 1997, 5). Dazzled by Ricardo's self-confidence and brilliance, Malthus seemed all too willing to serve as his sounding board.

Perhaps the most significant divide, and again one that has not previously been emphasized by scholars, was that of theology. Ricardo seems completely secular in his economic writings and correspondence. There is no appeal to a providential order, let alone a theological account of evil. Born and raised in the Jewish faith, Ricardo married a Quaker, an act that brought about a break with his religious heritage and a greater assimilation into mainstream English society. After amassing a formidable fortune as a stockbroker, he opted for a life of scholarship and then politics. As a member of Parliament for five years, he was active in promoting a number of reforms, including the religious toleration of Catholics and Jews. In private correspondence, Ricardo even ventured to acknowledge the rights of atheists (Milgate and Stimson 1991, 86).

John Pullen, Boyd Hilton, and Anthony Waterman have highlighted the theological component of Malthus's work, placing him within a broader tradition of Christian political economy (see Pullen 1981; Hilton 1988; and Waterman 1991). To this tradition also belong William Paley, Thomas Chalmers, Richard Whately, and, to a lesser extent, Richard Jones and William Whewell. Needless to say, each of these men had his own notions about what it meant to formulate a Christian political economy. But the primary question that they all addressed was the effort to reconcile the problem of human suffering with the notion of a benevolent deity.

Malthus's views evolved considerably, from the preacher-like two chapters on good and evil in the first edition of the *Essay*, to the hard-hearted remark about the limited seating at nature's feast in the 1803 edition, to the strong advocacy of moral restraint and the education of workers in the later editions.[14]

There is some debate about Malthus's allegiance to Christianity and likely conversion to utilitarianism. Daniel Malthus, his father, had befriended Bentham and hosted him in his home, so Malthus had possibly drunk in those ideas with his mother's milk. Certainly, it seems that he was unorthodox in his Christianity, at least on the questions of evil and the afterlife. And, despite his youthful tenure as a pastor for the Church of England, there is evidence that he had already begun to swerve from this calling while a student at Cambridge (see Pullen 1981, 45–47; and Waterman 1991, 96).

Several scholars, reaching back to Halévy, have deemed Malthus a *theological utilitarian,* a label that puts him in close company with William Paley (see Halévy 1972, 246; Harvey-Phillips 1984, 605; and Hollander 1997, 917).[15] The two men had a high regard for one another, and there is some evidence that their works were mutually reinforcing. Moreover, it seems that Malthus became more of a utilitarian with time, injecting additional phrases regarding the aggregate welfare of the population into successive editions of the *Essay,* and evincing a greater acceptance of the ongoing presence of vice and suffering. But there are still numerous indications that his theological roots were not forgotten, and, for that reason, I cannot endorse Hollander's view that "as social reformers Malthus and [John Stuart] Mill stand side by side."[16] Malthus was far more conservative, and far less willing to concede the malleability of individual character, than was Mill. Whereas Malthus endorsed the idea of a lottery of life characterized by deeply rooted injustices, Mill encouraged us to experiment with our lives, to take risks and to break from the mold. Moreover, Malthus's utilitarian remarks tended to be directed at the level of aggregate happiness, rather than at the level of individual acts. This again underscores the transition between the classical and the neoclassical, at which point individual deliberation and responsibility came to the fore.

That said, Malthus took a more tolerant position toward religious minorities than did many of his contemporaries and did not give much weight to heaven and hell as forces capable of curtailing human action. In some respects, he sounds quite progressive and liberal for his time, as Halévy, among others, has underscored. But on most political positions he was conservative, at least during his most active years. His long-lasting support for agrarian protection allied him with the Tories, although, over time, he appears to have become more Whiggish in his political allegiances.[17]

All these progressive tendencies notwithstanding, a reading of the *Essay* leaves the distinct impression that we live in a world created and governed by a Christian God. As Waterman has remarked, such convictions were in the very air that intellectuals such as Malthus breathed; to think otherwise was still heretical, at least in Britain, as the reaction to Darwin's *Origin of Species* amply demonstrated (Waterman 1991, 61; see also Brooke 2003). Hence the reaction to Ricardo, who was viewed as a harbinger of socialism and a more secular age.

While the theological underpinnings of the *Essay* are unmistakable, Malthus did not support his view by citing chapter and verse, as did other Christian political economists. For example, when arguing against the possibility of prolonging the human life span, he did not cite the many examples in the Bible of long-lived individuals (e.g., Methuselah). This seems part and parcel of his preference to look to the "Book of Nature" rather than to Scripture as the source of scientific inquiry and accords with his admonition to resist trying to fathom God's plan, a point that, as we saw, he made quite explicitly in the *Essay* but failed to follow himself.

DAVID RICARDO

Ricardo's main text, *The Principles of Political Economy,* also takes Smith as the point of departure, but it is a Smith without *The Theory of Moral Sentiments,* without the broader theological and philosophical concerns. There is almost no attention paid in the *Principles* to human motivation or the historical evolution of commerce and trade. With Malthus, however, there is a much more evident sense of the unfolding of a divine order. Indeed, the entire problem of human existence revolved around the reconciliation of the contradiction seemingly inherent in God implanting passions in humankind and then decreeing that suffering would ensue if those passions were given full vent. With Ricardo, there are no longer any passions. Political economy is no longer a tale of virtue and vice, at least not explicitly. True, landlords are implicitly parasitic insofar as they make no direct contribution to economic wealth. Still, Ricardo (himself a wealthy squire) made no pleas for the abolition of private property. Rather, the potential injustice of landownership did not figure in his political economy because he realized that, analytically, one could treat prices as the sum of the costs of labor and capital and ignore the return to land. In its "scientific," noncolloquial sense, *rent* was a return to the natural properties of the soil and was, thus, not the result of human negotiation. It rose over time, but this was due to the cultivation of marginal land, which in turn was the result of the natural path of population growth.

Wages and profits at least in the aggregate were determined by the bounties of nature, that is, by the annual harvest. Possibly the only important sphere of human life in which choice played a part for Ricardo was that of family size, which, again, was less the result of passion and more the result of economic well-being. Contrary to Malthus, Ricardo believed that population growth would diminish when workers raised their desire for goods: "The friends of humanity cannot but wish that in all countries the labouring classes should have a taste for comforts and enjoyments, and that they should be stimulated by all legal means in their exertions to procure them. There cannot be a better security against a superabundant population" (Ricardo 1817/1951, 100). There was no talk of exercising moral restraint, as in Malthus.[18] Ricardo granted laborers the right to struggle and protest against the owners of stock, to resist mechanization if it rendered them permanently unemployed, and to fight for suffrage and, thus, acquire a voice in Parliament.

Ricardo explicitly addresses the role of nature in economic production, taking issue with both Smith and Say. Like them, he sees land, air, and water as "gifts from nature." And, like them, he considers land to be different in that its supply is finite and, more important, access to it can be limited to those with might. But, in contrast to Smith, he grants nature a role in manufacturing. Ricardo quotes Smith at length in order to show his opposition. As we saw in the previous chapter, for Smith, when it comes to manufacturing, "*nature does nothing; man does all.*" For Ricardo, however, nature's contributions are paramount: "Does nature nothing for man in manufactures? Are the powers of wind and water, which move our machinery, and assist navigation, nothing? The pressure of the atmosphere and the elasticity of steam, which enable us to work the most stupendous engines—are they not the gifts of nature? to say nothing of the effects of the matter of heat in softening and melting metals, of the decomposition of the atmosphere in the process of dyeing and fermentation. There is not a manufacture which can be mentioned, in which nature does not give her assistance to man, and give it too, generously and gratuitously" (Ricardo 1817/1951, 76n). Thus, Ricardo's explicit emphasis on nature's role in economic production seems to be a result of an effort to reflect the growing importance of the manufacturing sector.

All this, of course, is a prelude to Ricardo's trump card, namely, the identification of rent as an unjust institution, an unwarranted payment to a minority of men for something given by nature. Nature's efforts to assist us "generously and gratuitously" are never ending, but some people are contingently fortunate to have others pay them for the use of the "original and indestructible powers of the soil." Indeed, Ricardo speculates: "If air, water, the elasticity of steam, and the pressure of the atmosphere, were of various qualities; if they could be appropriated, and each quality existed only in moderate abundance, they, as well as the land, would afford a rent, as the successive qualities were brought into use. With every worse quality employed, the value of the commodities in the manufacture of which they were used, would rise, because equal quantities of labour would be less productive. Man would do more by the sweat of his brow, and nature perform less" (Ricardo 1817/1951, 75). In other words, the emerging institution of rent is a strong indicator that man is getting less from the irreplaceable gifts of nature.

We will see that Mill finds nonsensical the idea of nature doing more or less in economic production. Ricardo's efforts to redeem nature in manufacturing from Smith's total spurning can, thus, be seen as an intermediary step toward Mill's belief that manufacturing and agriculture are on the same plane when it comes to nature's gifts. Mill, seeing Ricardo break through the conceptual barrier imposed by Smith, took the matter to its logical conclusion. Interestingly, while Ricardo appears to have endorsed Smith's view that nature plays a measurable role in agriculture, agreeing that the work of nature "is seldom less than a fourth, and frequently more than a third of the whole produce" (Ricardo 1817/1951, 76), he never estimated a comparable rate for manufacturing. It is simply more than zero.

Ricardo maintained in numerous passages that his theory of rent was the most essential brick in his entire economic edifice. As he instructed readers of the second and third editions of the *Principles:* "Clearly understanding this principle [that of rent] is, I am persuaded, of the utmost importance to the science of political economy" (Ricardo 1817/1951, 77). The reason is that rent could be eliminated from the theory of prices and distribution, thus, centered primarily on the conflict between capitalists and workers. After all, wages and profits were a just return on human effort. Rent was the result of an artificial institution that would not have existed if inferior lands had not been brought into cultivation.

This then gives added weight to Ricardo's concept of a *natural price,* which at first glance appears to be the normal prevailing price. For Ricardo, however, the natural price might never be actualized. It is composed of the natural wage and profit. By this he means less the fluctuations of the market, as Smith had maintained (the result of "higgling and bargaining" [Smith 1776/1976, 1:49]), than the returns to labor and capital that would prevail in a regime of no rent (a state of nature).

Ricardo's efforts to locate an immutable measure of value in the money price of gold were, ultimately, unsuccessful—and justly criticized by his contemporaries. What is odd, perhaps, is his subtle shift from taking gold as a hypothetical standard to one that is virtually legitimate. This shift suggests that he sought some natural base and believed, perhaps, that the quantity of gold in the earth's crust was, in some sense, the anchor of human economic value. For, while he also maintained that the natural price of gold was determined, not by supply and demand, but by the cost of production, at the end of the day he does resort to the fact of its limited supply. Nature determines the measure of value after all, insofar as it has only provided us with just so much gold.

Trade is another area in which nature seems to figure in Ricardo's account, notably the principle of comparative advantage. Ricardo highlighted the importance of natural endowments and local climate in determining what each region of the world would produce most economically. Were there no restrictions imposed on domestic and international trade or currency exchange, the world would achieve its most efficient and beneficial state:

> Under a system of perfectly free commerce, each country naturally devotes its capital and labour to such employments as are most beneficial to each. This pursuit of individual advantage is admirably connected with the universal good of the whole. By stimulating industry, by rewarding ingenuity, and by using most efficaciously the peculiar powers bestowed by nature, it distributes labour most effectively and most economically: . . . it diffuses general benefit, and binds together by one common tie of interest and intercourse, the universal society of nations throughout the civilized world. It is this principle which determines that wine shall be made in France and Portugal, that corn shall be grown in America and Poland, and that hardware and other goods shall be manufactured in England. (Ricardo 1817/1951, 133–34)

Each region, then, is designed to specialize in that good for which it has a comparative advantage. Although Ricardo makes no references to Providence, there seems to be a strong element of design in his conception of this global harmony, even if there is artifice involved. For what was new in Ricardo was the counterintuitive proposal of complete specialization. Even if Portugal can produce both wine and cloth more cheaply than England, each country is better off producing only the one good in which the advantage is comparatively greater. This is a far cry from the Linnaean picture of autarkic economics without specialization. But, in a sense, Ricardo's plan was just as much an appeal to nature's bounty. Recall that Linnaeus would have us building greenhouses in Sweden to grow tea and cocoa, to restore the Garden of Eden to an otherwise intemperate climate. For Ricardo, the entire world would become one large Edenic garden, with each nation cultivating that which nature most intended.

Some qualification is called for here, however, since Ricardo also invited nations to pursue their manufacturing comparative advantages even if, as in the case of cotton production in Britain, their climatic disadvantages suggested that they focus their efforts elsewhere. In other words, there was no geographic determinism in Ricardo's vision of nation-specific production. Those countries with more advanced industrial sectors and greater technical know-how would be wise to produce and export manufactured goods, whatever their natural endowments. In that sense, natural endowments may figure only indirectly in Ricardo's vision for the utopian global economy.

Nevertheless, Ricardo, like his predecessors, still privileges agriculture over manufacturing. The latter is much more susceptible to "the influence of fashion, prejudice, or caprice," and it is also more vulnerable to wars, new taxes, and other afflictions (Ricardo 1817/1951, 263). Hence, the demand for any particular manufactured good can fluctuate considerably. Ricardo grants that agriculture is also subject to "contingencies of this kind, though in an inferior degree," precisely because the demand for food is ubiquitous and "uniform." But, insofar as only wealthy and fully developed nations will become those who manufacture goods for others, they are in a much better position to withstand these vicissitudes. Ricardo deems such fluctuations to be "an evil to which a rich nation must submit; and it would not be more reasonable to complain of it, than it would be in a rich merchant to lament that his ship was exposed to the dangers of the sea, whilst his poor neighbour's cottage was safe from all such hazard" (266).

The reason that such complaints are unwarranted is that nations always recover from such "retrograde states." Not only is there a tendency to rectify economic ills, but, once an economy is thriving, it may be sustained for an indefinite period. Unlike our own passage from youth to adulthood and then to the grave, Ricardo grants nations a "natural tendency to continue for ages, to sustain undiminished their wealth, and their population" (Ricardo 1817/1951, 265). So here he is no longer using the commonplace Enlightenment analogy that nations are organic beings subject to life cycles. They do not necessarily decay, as Hume had supposed. Nor does Ricardo endorse Smith's famous remark about the "narrow capacity of the human stomach," used to argue for limits on the amount of capital required by agriculture. Ricardo emphasizes the potential for additional capital accumulation, both in the agrarian and in the manufacturing sectors.

Ricardo and Malthus regarded political economy as a science, less exact than physics or astronomy to be sure, but a science nonetheless.[19] They therefore sought out general principles on the assumption that the world was governed by a limited set of laws. The question was to what extent these laws were grounded in the physical properties of the world and to what extent they were the result of human agency. For Malthus, physical constituents generally prevailed over institutional ones. Wages, for example, were determined by the ratio of the population to the harvest each year. Even commodity prices were determined by physical quantities, by the cost of production, more than by the institutional features of markets (see Ricardo 1817/1951, 382). Human industry and population growth were also subject to the physical constraints imposed by the principle of diminishing returns. The stationary state that loomed on the horizon was one defined by nature, not by human agency. Even the one key decision left to humankind, the rate of reproduction, was more subject to passion than to reason.

Malthus also tended to view the economy in terms of shorter periods of turnover, in keeping with his agrarian predilections. As Ricardo noted in a letter of 1817 to Malthus:

> It appears to me that one great cause of our difference in opinion, on the subjects we have so often discussed, is that you have always in your mind the immediate and temporary effects of particular changes—whereas I put these immediate and temporary effects quite aside, and fix my whole attention on the permanent state of things which will result from them. Perhaps you estimate these temporary effects too highly, whilst I am too much disposed to undervalue them. To manage the subject quite right they should be carefully distinguished and mentioned, and the due effects ascribed to each. (Ricardo 1951–73, 7:120)

Ricardo found the question of capital accumulation, both in the agrarian and in the manufacturing sectors, more absorbing than did Malthus. As he remarked in an 1815 letter to Malthus: "By facility of production I do not mean to consider the productiveness of the soil only, but the skill, machinery, and labour joined to the natural fertility of the earth" (Ricardo 1951–73, 6:292). The very fact that he had to articulate his appreciation of factors other than the soil is highly suggestive of the differences between them. Ricardo was also much more attentive to the profit rate, a phenomenon that was much more evident in the commercial and manufacturing sectors than in the agrarian. Overall, it seems that Ricardo had come to place greater emphasis on human agency and institutions than had Malthus.

Keith Tribe has suggested that, in Ricardo, *capital* acquired a new meaning (Tribe 1978, 141). It was seen, no longer in the eighteenth-century sense as equivalent to personal wealth, but, rather, as an advance to further production. In that sense, it had no obvious limit. Malthus granted that the accumulation of capital had no limit, but he still tied it to individual wealth. In a letter to Ricardo, he remarked: "I have always allowed that the progress of capital and population, while they can go on together, uninterrupted by the difficulty of procuring subsistence, is absolutely unlimited; but I most distinctly deny that the demand for capital is unlimited, with a limited population; and this appears to me to be the proposition that you maintain" (Ricardo 1951–73, 6:318). Malthus clearly did not grasp the fact that demand is always potentially unlimited and not a function of population.

Tribe has also maintained that, for Ricardo, rent was more about forms of capital than about the natural powers of the soil. I agree with Tribe that the formation and evolution of a schedule of rents in the Ricardian sense is about an economy undergoing capital accumulation and that this transpires independently of individual choice, or what he calls "the conscious activity of the economic agents concerned" (Tribe 1978, 120). Capital accumulates as a result of inexorable laws of economic growth and development, coupled with population growth and the principle of diminishing returns, which forces humans to become enterprising. There is, in short, little scope for individual economic agency in Ricardo. Because his economic analysis is grounded on the three factors of production—land, labor, and capital—and the groups of persons associated with each factor—landlords, workers, and capitalists—Ricardo's conception of the economy is still more allied with eighteenth-century political economy than with neoclassical economics of the 1870s, which tended to downplay these categories and render them perfect substitutes. Indeed, for neoclassical economists, land tends to vanish altogether. Among British economists, Jevons's *Coal Question* (1865) marks the point at which economists turned away from natural resources. It is highly ironic that one of the vanguard of the neoclassical school would still use strong Malthusian analysis (see Peart 1996, chap. 2).[20] Yet, by drawing attention to Britain's energy needs, Jevons also helped put land, labor, and capital on a level playing field. The fuel that runs the steam engines was, indeed, the great catalyst for rendering all the factors into perfect substitutes. But it is the absence of individual deliberation that most separates the classical from the neoclassical theory. In that sense, again, the economy of Ricardo's day, at least on paper, was more governed by physical nature than by human nature.

For Ricardo, the three classes and their conflicts were as fixed and immutable as Linnaean species. Moreover, they were bound together by market processes in ways that were inextricable. It was in this sense, of an integrated set of relations, that Ricardo cultivated the concept of an economy. Linnaeus's natural oeconomy had finally transmuted into a human economy, albeit with physical nature etched in the foreground.

CLASSICAL POLITICAL ECONOMY AFTER RICARDO

It is not possible to cover the life and ideas of all those economists—several dozen, if one includes popular figures such as Robert Owen and Walter Bagehot in addition to those affiliated with universities and scientific societies— active in the fifty-year period after Ricardo's *Principles* appeared. The most prominent theorists in the years before the appearance of J. S. Mill's *Principles of Political Economy* (1848) were John Ramsey McCulloch and Nassau Senior. But Thomas Tooke, Richard Whateley, Jane Marcet, and Harriet Martineau were more widely read. Numerous others published on monetary questions or the theory of value, notably Mountiford Longfield, Samuel Lloyd, Samuel Bailey, Thomas De Quincey, Robert Torrens, and the Earl of Lauderdale, to name just a handful. There was also a group of Ricardian socialists that included John Francis Bray, Thomas Hodgskin, William Thompson, and, arguably, Karl Marx, although his work and influence came considerably later. At Cambridge, William Whewell inspired a group of mathematical economists, the best known of which was Dionysius Lardner. Whewell also encouraged Richard Jones to cultivate empirical studies and, thus, began a school of economic historians—the most notable being Arnold Toynbee—which attempted to counter mainstream theorists (see Collini, Winch, and Burrow 1983, 247–76).

I will visit the work of McCulloch and Senior briefly because they are the most orthodox and prominent classical economists of the period before Mill. They held the most distinguished academic posts (London and Oxford, respectively), issued prominent publications, and advised the government. I will focus on the extent to which the economy was still seen as part of physical nature and human agency now perceived as a major component of economic properties.

Although McCulloch is often considered to be an arch-Ricardian, more Ricardian than Ricardo himself, I think that Denis O'Brien is correct in emphasizing that McCulloch had moved quite a ways away from his mentor (O'Brien 1970, 121–22). True, he has nothing but praise for Ricardo's powerful mind and disdain, if not scorn, for Malthus's failure to understand it, but the thrust of his own *Principles of Political Economy* (1825) is considerably different than that of Ricardo's *Principles,* certainly much more so than Mill's later *Principles.*

Mary Poovey has argued more recently that, in his efforts to popularize political economy, McCulloch restored the veneer of providentialism that had waned in the hands of Ricardo, but he did this with appeals to nature, not to divine planning: "McCulloch also wanted to make the conclusions that liberal political economists had reached seem like discoveries of natural laws, not impositions of some a priori theory" (Poovey 1998, 296).

Most of the numerous *Principles* to come after Ricardo are significantly inferior in quality, at least in terms of rigorous analysis. McCulloch's own is no exception. But it has its own strengths. His account is much more discursive, fleshing out Ricardo's frame. While the first edition was streamlined for publication in the *Encyclopedia Britannica,* by the third and much-expanded edition of 1842 we have ideas that are more in line with Mill than with Ricardo.[21]

For one thing, McCulloch sheds the productive/unproductive distinction that had been bandied about in Smith, Malthus, and Ricardo to the point of utter confusion. His argument is not particularly sound, however. He does not deny the frivolity of a stage actor. But, insofar as we are willing to pay to attend theatrical events, we are prompted to work. Hence: "What is a cause of production must be productive" (McCulloch 1864/1965, 507). Moreover, for agriculture to be productive we must have the security of property and, thus, judicial and military systems. By analogy, a judge or a soldier is just as productive as a shepherd: "If the herdsmen who protect a single corn-field from the neighbouring crows and cattle be productive, then surely the judges and magistrates, the soldiers and sailors, who protect every field in the empire . . . have a good claim to be classed among those who are super-eminently productive" (509). With similar arguments, McCulloch also dissolved the distinction between productive and unproductive consumption. Notice that the temporal duration of the activity is no longer a criterion. Indeed, George Poulett Scrope criticized McCulloch for classifying billiard playing and blowing soap bubbles as productive activity (Scrope 1833/1969, 46n). McCulloch did not mention these particular pursuits, but that they are productive activities follows from his definition in that they could be construed as contributing to our willingness to be industrious.

McCulloch also grants a larger role to human agency. He discusses the division of labor, not as something that unfolds of its own accord, as Smith had depicted it, but as something consciously driven by human decision-making. He also extolls the distinctively human trait of reason: "It is the proud distinction of the human race, that their conduct is determined by reason, which, though limited and fallible, is susceptible of indefinite improvement. Man is destined to be the artificer of his own fortune" (McCulloch

1864/1965, 21). McCulloch thus maintains that the potential for mechanization is unbounded, expressing great optimism in the progress of scientific inquiry (68–69). We have here the image of "pressing the powers of nature into our service," rather than succumbing to her inexorable forces (51). As a result, the fact that agriculture is no more tied to nature than is manufacturing is underscored. Both are the result of human ingenuity, human tools, and human practices. For that reason, McCulloch believes that previous economists have been misled into giving higher priority to agriculture than to manufacturing and commerce. At root, they are the same.

McCulloch directly addresses Smith's claim that nature offers no assistance in manufacturing, in contrast to the generous hand given agriculture. Turning the proposition on its head, McCulloch suggests: "There are no limits to the bounty of nature in manufactures; but there are limits, and those not very remote, to her bounty in agriculture" (McCulloch 1864/1965, 121). The elasticity of steam, the malleability of metals, and the processes of bleaching and fermentation, all are the work of nature.

As a result of this conceptual breakthrough—one that Mill will also make—McCulloch maintains that profits are simply "a consequence of the bounty of Nature" (McCulloch 1864/1965, 65). They are due, not to specific ingenuity, or to the level of the division of labor, but to the difference between that which is produced in a given period and that which is consumed. Exchange is always conducted under the banner of equality, with the inexorability of a law of nature: "Buying and selling are in commerce, what action and reaction are in physics, always equal and contrary" (McCulloch 1825, 135). McCulloch thus appears to have in mind something more akin to our notion of a normal profit under perfect competition. There are no arbitrageurs or monopolists in the picture he paints.

McCulloch's realization that manufacturing is as much tied to nature's bounty as is agriculture also led him to offer a different definition of *capital,* one that sparked considerable debate. For McCulloch, wine aging in the cellar was accumulating capital, even if the only work being done was the natural process of fermentation. The process is really no different from corn growing in the soil. The change in the good is brought about entirely by the operation of nature. Nevertheless, this is, not the source of the profit, but, rather, the return on the efforts of the wine merchant to see that the grapes are cultivated, pressed, and stored in caskets. The profit is really a return on that labor and capital (McCulloch 1864/1965, 293–94).

McCulloch has clearly gone full circle here since he earlier insisted that all profits are due solely to the bounty of nature. He has committed the common error of confusing aggregates with individual cases. The enterprise of making the wine and storing it for several years justifies the profit, the return to the producer and merchant. Otherwise, they would employ their capital in other sectors. But, in the aggregate, in the case of equality through exchange and, implicitly, perfect competition, profit is due solely to the excess of price over costs.

Nassau Senior, the first Drummond Professor at Oxford, is often acknowledged for placing more emphasis than his predecessors on the role of utility in the analysis of prices and wealth and for injecting the idea of abstinence into the theory of capital accumulation. But he also addresses the contributions of nature and draws some distinctions among the views of Smith, Ricardo, and McCulloch. For one thing, he adopts the term *natural agents* in place of *land*, which is but the most common species of a genus that also includes waterways and mines (see Senior 1836/1965, 90). How interesting that Senior has granted to nature agency, a property normally reserved for rational beings. To be consistent, he also refers to the *proprietors of natural agents* rather than to *landowners*.

Senior's use of the term *agent* is not accidental. In his analyses of rent and of the distinction between a commodity and a service, he addresses explicitly the relative role of the agency of nature vis-à-vis that of the agency of man. Whereas profits and wages are exclusively a return to human agency—to abstinence and labor—rent is a return solely to nature's owners. It is "a reward . . . for not having withheld what he was able to withhold, for having permitted the gifts of nature to be accepted" (Senior 1836/1965, 90).

Moreover, it is not that nature is so parsimonious in agriculture, as Ricardo had maintained, but that it is generous. Indeed, rents have risen secularly precisely because nature is bountiful: "The condition essential to the payment of the labour of nature is not, as Mr. Ricardo states it, that her assistance shall be little, but that it shall not be infinite." There is no ceiling to rents. For rent depends on neither the exertions of the proprietor nor the will of the leasee. "It is all pure gain," the result of competitive market conditions (Senior 1836/ 1965, 139).

Senior also takes a look at the global economy. Without mentioning Providence, he nonetheless points to a grand scheme for a harmonious order: "Nature seems to have intended that mutual dependence should unite all the inhabitants of the earth into one commercial family. For this purpose she has indefinitely diversified her own products in every climate and in almost every extensive district. For this purpose, also, she seems to have varied so extensively the wants and the productive powers of the different races of men" (Senior 1836/1965, 76). Presumably, as an agent, Senior has no qualms granting nature intentionality as well.

In another case, Senior contrasts an English fisher of oysters, who by chance finds a pearl, with a pearl fisher in Ceylon, who is unequivocally a producer of pearls. In both cases, the pearls are the result of the "agency of nature," but only in the Ceylonese case is human agency significant in the production of the pearl. In England, the fisherman is a producer solely of oysters and a casual finder of pearls (Senior 1836/1965, 53).

Once again, Senior emphasizes the role of intentions that precede actions. He thus opens the door to human rationality as a point of departure for economic theory. Foresight, intentionality, is the key to distinguishing true economic production from mere activity. *Abstinence,* Senior's preferred term for the use of capital in production, is, of course, a variation on the same theme. It again brings out the role of agency, of self-denial, and of a preference for distant rather than immediate gain (Senior 1836/1965, 58–60). Senior notes that he chose the term with care, over *Providence* and *frugality* (59), since he wanted to underscore the idea of sacrifice without explicit labor or effort.

Both McCulloch and Senior draw numerous references to nature, nature's intentions, and nature's gifts, and, in the case of the later editions of McCulloch's *Principles,* Providence itself makes an appearance more than once. Senior, whom Boyd Hilton maintains was thoroughly secular, evades the term, but there are still appeals to natural design and order (Hilton 1988, 45). Most others at the time still clung to the idea of a natural order. George Poulett Scrope, for example, highlighted the role of natural laws in his *Principles of Political Economy* (1833). Richard Whately spoke freely of a providential wisdom whereby rich and poor contribute to one another in a mutually beneficial manner (Whately 1832, 99–100; Hilton 1988, 52).

The leitmotiv of classical economics, then, pointed to the physical constraints of nature setting limits to human prosperity. Human agency was still subordinate to the forces of nature, although there were glimmerings of paths for conquest. We will see with John Stuart Mill that man can come to be almost fully in charge of his destiny, although wealth is still bounded. With the neoclassical theory that emerged in the 1870s and 1880s, wealth was cast in terms of mental satisfaction or utility and, thereby, become truly unlimited.

CHAPTER SEVEN

Mill and the Early Neoclassical Economists

It is a case of the error too common in political economy, of not distinguishing between necessities arising from the nature of things, and those created by social arrangements: an error, which appears to me to be at all times producing two opposite mischiefs; on the one hand, causing political economists to class the merely temporary truths of their subject among its permanent and universal laws; and on the other, leading many persons to mistake the permanent laws of Production.

—John Stuart Mill, *Principles of Political Economy*

John Stuart Mill's *Principles of Political Economy* (1848) served as economists' primary text for over forty years and helped resolve many of the controversies that stemmed from the writings of Malthus and Ricardo.[1] Mill was also the first to write extensively on methodology. Others before him had declared political economy to be a science, but only with Mill's celebrated 1836 essay "On the Definition of Political Economy and on the Method of Investigation Proper to It" and his later *System of Logic* (1843) do we find a detailed treatment of the subject. Much has now been written on Mill's methodological pronouncements as well as on the more widespread effort by economists to emulate Newtonian mechanics.[2] Two prominent philosophers, Nancy Cartwright (1983) and Daniel Hausman (1992), have rekindled an appreciation for Mill's grasp of the limitations of scientific explanation. But there is another aspect of Mill's political economy that, I believe, has not been given the attention it deserves. Mill had a different conception of the economic order than his predecessors did. His conception of the economic realm was more detached from physical nature, more contingent on human institutions, and, thus, more artificial. Mill sheds the more traditional set of appeals to physical nature and, thus, recognizes the economic world as the product of human agency and institutions that are quite at odds with nature.

In the 1836 "On the Definition of Political Economy," Mill addresses the problem of demarcating political economy from the physical sciences.[3] To study economic production is necessarily to study physiology, chemistry, mechanics, and geology. Thus, political economy "presupposes all the physical sciences" in its inquiry. Moreover, as Mill notes, "there are no phenomena which depend exclusively upon the laws of mind." Hence: "The real distinction between political economy and physical science must be sought in something deeper than the nature of the subject matter; which, indeed, is for the most part common to both." Because the "laws of mind and [the] laws of matter are so dissimilar in their nature," however, it is possible to separate our study of the two. Political economy, while encompassing all the physical sciences, may conveniently take as its domain of inquiry "the phenomena of mind which are concerned in the production and distribution of those same objects" (Mill 1836/1967, 316–18). Here, then, is, not only an explicit recognition of the divide between the material and the mental, but also an expressed belief that the proper area of inquiry for the economist is the mental, with perhaps an occasional glance at the material background.

When it comes to the *Principles of Political Economy*, however, it is not clear that Mill is guided by these assertions. For one thing, he rules out the possibility of laws of consumption, even though they would appear to be prime candidates for mental laws (see Mill 1967, 318n).[4] More to the point, Mill begins his text with the laws of production precisely because they are the most fundamental in a material sense: "The laws and conditions of the production of wealth partake of the character of physical truths. There is nothing optional or arbitrary in them. Whatever mankind produce, must be produced in the modes, and under the conditions, imposed by the constitution of external things, and by the inherent properties of their own bodily and mental structure" (Mill 1871/1965, 2:199). Elsewhere, he comments: "The conditions and laws of Production would be the same as they are, if the arrangements of society did not depend on Exchange, or did not admit of it" (3:455). In short, the laws of production transcend any specific social conditions. They arise from the "nature of things" and are, therefore, permanent truths, a prime desideratum in the construction of good science.

But Mill aligns his concept of economic production with physical nature even more explicitly, and he does so from the opening pages of the *Principles*.

Nature and only nature is truly productive, not only because it provides all the objects for economic production, but also because it provides the requisite forces of nature: "Nature, however, does more than supply materials; she also supplies powers." In fact, nature supplies all the power. Mill considers the case of a producer of linen cloth. It appears that the spinner did everything by hand, "no natural force being supposed to have acted in concert with him." But, in fact, what held the fibers together is the result of the force of cohesion, a force of nature. Mill then maintains: "If we examine any other case of what is called the action of man upon nature, we shall find in like manner that the powers of nature, or in other words the properties of matter, *do all the work*" (emphasis added). Human labor simply puts certain objects into appropriate configurations; "the properties of matter, the laws of nature, do the rest" (Mill 1871/1965, 26–28).[5]

Moreover, Mill insists, our ability to harness the powers of nature does not come in degrees. It is not the case that nature contributes more to the efficacy of human labor in one occupation than it does to that in another: "The part which nature has in any work of man, is indefinite and incommensurable. It is impossible to decide that in any one thing nature does more than in any other." As a result, agriculture is no closer to nature in the raw than is manufacturing. Mill explicitly impugns the position taken by the Physiocrats and Adam Smith, that is, "supposing that nature lends more assistance to human endeavours in agriculture, than in manufactures" (Mill 1871/1965, 26–29).[6] In utilitarian terms, manufacturing is on a par with agriculture. "How is it possible," Mill asks, "to say whether agriculture, or manufactures, be most productive of wealth? unless it is pretended to determine whether food or clothing be most essential to the happiness of man" (Mill 1967, 4:42). He has implicitly erased the traditional appeal to the historical priority of agriculture and its associations with man in a state of nature.

Human labor is conceived of in much more plastic terms: "All the labour of all the human beings in the world could not produce one particle of matter. . . . What we produce, or desire to produce, is always . . . utility. Labour is not creative of objects, but of utilities" (Mill 1871/1965, 46). Mill's appeals to utility, and, hence, to a nonphysical conception of wealth, enabled him to reinforce the view that political economy is essentially a mental rather than a material science. In this respect, then, he has taken a sizable step toward placing the economic order within human hands.

Yet Mill seems reluctant to let go of the concrete and physical altogether. In one respect, it seems, utility is itself something substantive. This is apparent in his reaffirmation of the long-standing distinction between productive and unproductive labor. Smith had drawn the distinction in terms of the potential for labor to crystallize in an object. Mill recasts the distinction in terms of utility, but it is utility that can be "fixed" in objects and extracted at a later point in time. To take his own example, the violinmaker and the violin teacher are both engaged in productive labor; the first transforms wood and horsehair into a permanent object, and the second imprints on pupils' minds the requisite technical skills. But the concert violinist is unproductive (Mill 1836/1967, 285). Whereas, in the first two cases, one can extract utility at a later point in time, in the third the utility produced by the performer evaporates at the very point of its production. The early neoclassical economists were to discard this distinction as nonsensical. William Stanley Jevons, for example, pointed out that, if labor produces only utility, what difference does it make whether it leaves an imprint on an object (see Jevons 1906/1965, 85–89)? For Jevons, utility existed only at the instant of production. It had no sustaining features. Mill's conception of the distinction marks an appropriate point of transition between the Smithian and the Jevonian points of view.

Other aspects of Mill's conception of the economy suggest a predilection for features grounded in material nature rather than the mind. The principle of diminishing returns is deemed to be "the most important proposition in political economy" in that, as Mill frequently notes, it demonstrates the physical limitations of human industry (Mill 1871/1965, 2:173–74). Furthermore, in keeping with a long tradition, Mill tries to peel away the layer of money that coats economic activities. As he puts it: "There cannot . . . be intrinsically a more insignificant thing, in the economy of society, than money" (3:506). The law of wages, for example, would not be affected one whit by the introduction of money. Wages are governed by the ratio between population and capital and, thus, either directly or indirectly, by physical quantities.

MILL ON NATURE

Much more can be gleaned of Mill's view of man's place in nature by turning to his essay "On Nature," which was written in 1852–53 but published only posthumously, in 1874, as one of the three *Essays on Religion*.[7] These essays constituted an attack on natural theology and, thus, seem to have little bearing on economic analysis. Yet the distinction that Mill draws in them between the natural and the artificial has far-reaching consequences for his broader program for social reform and, hence, for his economic thought.

Mill notes that the word *nature* is laden with emotive power and that, over the ages, different people have made drastically different appeals to nature to serve their own ends. But, while an array of definitions of the term can be found, they all boil down to just two. Either nature is the sum of all phenomena, "not only all that happens, but all that is capable of happening," or it is that which "takes place without the agency, or without the voluntary and intentional agency of man" (Mill 1874/1969, 375).

Mill grants the plausibility of the first definition. But, he argues, it has the unfortunate consequence of depriving humans of moral responsibilities. "To bid people conform to the laws of nature," he writes, "when they have no power but what the laws of nature give them . . . is an absurdity." Nor could nature serve as a guide for human morality, for it is replete with torture and destruction:

> In sober truth, nearly all the things which men are hanged or imprisoned for doing to one another, are nature's every day performances.

> Anarchy and the Reign of Terror are overmatched in injustice, ruin, and death, by a hurricane and a pestilence. (Mill 1874/1969, 379, 385–86)

To look to nature as a model for human action would be to sanction some of the most evil of deeds.

Moreover, Mill argues, it would be highly arbitrary to seek the good in parts of nature and dismiss the rest as evil. Natural theologians, if they endorse active intervention, are thus engaged in contradictions: "If the natural course of things were perfectly right and satisfactory, to act at all would be a gratuitous meddling, which as it could not make things better, must make them worse." As part of his repudiation of natural theology, then, Mill argues for the second definition of nature. There is a distinct realm of the artificial that conflicts with nature. More emphatically, everything that man does is in opposition to nature. All the bridges, wells, and lightning rods that we construct "acknowledge that the ways of Nature are to be conquered, not obeyed" (Mill 1874/1969, 380–81).

Given this concept of nature, why, in his political economy, does Mill still privilege physical nature? Why does he characterize labor in such a passive way, as simply the harnessing of nature's powers? And why does he insist that production is independent of human institutions and, thus, governed by natural laws when in "On Nature" he sees all human agency as the confrontation of nature for the better? Was he simply inconsistent (the matter is, after all, difficult), or did he consciously decide not to revise his theory of political economy in the light of his later thoughts on the distinction between the natural and the artificial?

There is, I believe, no one, simple answer. Certainly, it is tempting to assume that the oeuvre of a writer as prolific as Mill will inevitably end up harboring serious contradictions. But there is only a five-year gap between the *Principles* and "On Nature." Moreover, Mill lived long enough to oversee seven editions of the *Principles* and even late in life proved himself capable of owning up to erroneous beliefs, such as his celebrated recantation of the wages-fund theorem (see Hollander 1985, 387–422). He thus had ample opportunity to revise his privileging of physical nature in the *Principles* if he so wished. Furthermore, it is reasonable to suppose that, on such a fundamental matter as the distinction between the natural and the artificial, Mill would have been consistent from one work to the next.

One explanation may be found in the recognition that, in the field of political economy, Mill labored under the weight of tradition. We know that, while versed in Ricardian economics at the age of thirteen, when it came to his main treatise on the subject, he put pen to paper only a few years before publication. In so many passages, Mill seems to be trying to break new ground with his conception of political economy as a mental science, but perhaps he was not always able to follow through on his intuitions. The physical threads woven into his *Principles* might be viewed as vestiges of prior economic thought that he was unable or disinclined to shed.

Given the opposition posed by Auguste Comte—to wit that political economy had no claim to be considered a separate science—Mill may have retained the physical trappings of the classical theory as a way of reinforcing the discipline's scientific stature. Certainly, he advanced detailed arguments for the methodological identity of political economy and Newtonian mechanics. Moreover, his conception of a separate and unique economic realm called for a fully developed science of the mind, something he lacked. Even his efforts to formulate a new discipline of ethology or character formation did not bear fruit (see Leary 1983). Rather than rock the very foundations of political economy, he may have been inclined to keep to the traditional approach.

How did the thorny problem of human nature fit into this distinction between physical nature and the social realm? There appears to be a deep tension in Mill. In both the *System of Logic* and the *Principles of Political Economy,* he commits himself to a significant degree of nomotheticity within human nature. As he remarks in the *Principles:* "The opinions and feelings of mankind, doubtless, are not a matter of chance. They are consequences of the fundamental laws of human nature, combined with the existing state of knowledge and experience, and the existing condition of social institutions and intellectual and moral culture" (Mill 1871/1965, 2:200).

Nevertheless, in "On Nature," Mill insists, not just that human nature is modifiable, but that it ought to be modified: "The duty of man is the same in respect to his own nature as in respect to the nature of all other things, namely not to follow but to amend it." Thus, he takes human nature to be anything but fixed, a view that sets him apart from his eighteenth-century predecessors. Indeed, he calls into question the belief that there are universal traits. Certain more unrefined sentiments, such as fear or selfishness, tend to be more widespread. But their counterparts, such as courage or sympathy, may be as prevalent in what Mill took to be the uncivilized regions of the globe as in Victorian England (Mill 1969, 397).[8] Virtue, Mill insists, comes only through effort and reform. Apparently, we have planted within us the capacity for such virtues as honesty, courage, and benevolence. More important, many of us have the intelligence to recognize virtue when it springs forth and, thus, "to cherish the good germs in one another." But the cultivation of the seeds of virtue is strictly artificial and not at all natural: "Whatever man does to improve his condition is in so much a censure and a thwarting of the spontaneous order of Nature." Hence, Mill contends: "This artificially created or at least artificially perfected nature of the best and noblest human beings, is the only nature which it is ever commendable to follow" (Mill 1874/1969, 381, 396–97).

Like many of his contemporaries, Mill subscribed to a belief in the uniformity of nature. Indeed, for him, everything was subject to one law or another. But he also had to leave room for social reform and the improvement of human character. The question, then, was where to draw the line between the necessary and the contingent—or, to put it another way, where to locate the area amenable to reform without granting that everything was subject to change.

The solution, at first glance at least, seems to lie in the distribution of wealth among the different social classes. We have already noted that production is akin to the laws of physics and, thus, nonarbitrary. The distribution of wealth, by contrast, "is a matter of human institution solely. The things once there, mankind, individually or collectively, can do with them as they like." Here is the region for genuine reform, particularly, Mill hopes, the reform of property rights. Yet, even within the realm of the distribution of wealth, laws take effect. As Mill puts it, we must "consider, not the causes, but the consequences, of the rules according to which wealth may be distributed. Those, at least, are as little arbitrary, and have as much the character of physical laws, as the laws of production. Human beings can control their own acts, but not the consequences of their acts either to themselves or to others" (Mill 1871/1965, 2:200). What is distributed involves human deliberation. But, once that deliberation has resulted in the issuing of decisions, everything else is determined.

Mill thus concedes more room for human agency, especially deliberation, than did his predecessors. His favorable disposition toward the doctrine of laissez-faire was not based on a belief that it was better for nature to take its course. On the contrary, there was nothing good or just in nature: "All praise of Civilization, of Art, of Contrivance, is so much dispraise of Nature; an admission of imperfection, which it is man's business, and merit, to be always endeavouring to correct or mitigate." Even granting the existence of God, Mill still argued forcefully that humans act best by confronting rather than conforming to nature: "If Nature and Man are both works of a Being of perfect goodness, that Being intended Nature as a scheme to be amended, not imitated, by Man" (Mill 1874/1969, 381, 391).

Previous economists, Ricardo most saliently, saw distribution as entirely subject to law. Given competitive conditions, profits, wages, and rents were completely determined. Mill recognizes this when he remarks: "Only through the principle of competition has political economy any pretension to the character of a science. So far as rents, profits, wages, prices are determined by competition, laws may be assigned for them . . . and scientific precision may be laid down." But, he emphasizes, there is never pure competition. Custom always enters in to one degree or another and is especially prevalent in the case of land tenure (Mill 1871/1965, 2:239–40). In this respect, Mill opened the door to more human deliberation and rulemaking.

Economic advancement consists in increasing our powers over nature. Mill is able to see around the Malthusian trap, in part because of his optimistic belief that humans will learn to exercise birth control and in part because of his appreciation of the forward march of industry. The two go hand in hand: "Poverty, like most social evils, exists because men follow their brute instincts without due consideration. But society is possible, precisely because man is not necessarily a brute. Civilization in every one of its aspects is a struggle against the animal instincts. Over some even of the strongest of them, it has shown itself capable of acquiring abundant control. It has artificialized large portions of mankind to such an extent, that of many of their most natural inclinations they have scarcely a vestige or a remembrance left" (Mill 1871/1965, 2:367).

The link between the natural order and justice is also severed by Mill. As he put it—quite bluntly—in "On Nature" (echoing Hume): "Justice is entirely of artificial origin" (Mill 1874/1969, 396). The just society will come to pass only when man has become fully civilized and cultivated the seeds of virtue. As Mill declares in the opening pages of the third edition of the *Principles:* "The great end of social improvement should be to fit mankind by cultivation, for a state of society combining the greatest personal freedom with that just distribution of the fruits of labour, which the present laws of property do not profess to aim at" (Mill 1871/1965, 2:xciii).

In many respects, Mill may be viewed as putting the capstone on the classical theory of political economy. But, in certain fundamental respects, he ushered in our current conception of a denaturalized economic realm. There is, to be sure, a sense in which economists still believe that they study a physical world, even though, when it comes to explanations of production, concrete definitions of capital and labor are riddled with inconsistencies. Expectations—the mental, not the material—drive the economy. Man is to be set apart from physical nature; the good and the just must be sought within the civilized world. The artificial is obviously much superior to the natural.

How paradoxical that this profound repositioning of the economic realm transpired at virtually the same point in time—the 1840s and 1850s—as naturalists such as Charles Darwin were purportedly economizing nature, reading supply and demand, division of labor, competition, and Malthusian pressures into the world of plants and animals (see Schweber 1985). Robert Young (1985b) has argued forcefully that the most central question of Victorian England was man's place in nature. But, while naturalists were bringing man into nature, treating him as just another biological species and reducing his intellectual and moral capacities to animal instincts, political economists such as Mill were taking him out.

Joseph Schumpeter once described Mill's *Principles* as a halfway house (Schumpeter 1954, 603). There is a great deal of truth to this remark, although that truth is, in a sense, different from that intended by Schumpeter. Schumpeter focused on Mill's analytic oscillations between Ricardian and neoclassical tenets. But it was, in fact, Mill's concept of the economy itself that took him a considerable step away from his predecessors. The economy was set apart from the natural order and seen as an instrument for the amelioration of humankind. Political economy itself was no longer a material science. Although it presupposes the operation of the laws of physiology, chemistry, mechanics, and so forth, it takes as its proper domain of inquiry mental phenomena. Mill thus paved the way for the subsequent declarations by the early neoclassical economists who grounded the subject so firmly in the mind.

THE PSYCHOLOGICAL TURN

Why did Mill make these shifts? The answer lies partly in his appreciation of psychology, initially fostered by his father. In 1859, in response to the work of Alexander Bain, he declared: "The sceptre of psychology has decidedly returned to this island. The scientific study of mind, which for two generations, in many other respects distinguished for intellectual activity, had, while brilliantly cultivated elsewhere, been neglected by our countrymen, is now nowhere prosecuted with so much vigour and success as in Great Britain" (Mill 1859/1978, 341).

During the Victorian era, the conception of the economy and its salient features underwent a significant transformation. One critical factor in this process was an unprecedented but relatively short-lived enthusiasm among economists for psychology, starting with Mill's declaration of 1859, and ending more or less with Marshall, who died in 1924.

Mill's enthusiasm for psychology, or the science of the mind, must be seen in the context of a larger movement, one that he helped spearhead. The key figure in this movement was Alexander Bain, whose *The Senses and the Intellect* (1855) did much to wed physiology to the associationist psychology of David Hartley and James Mill. Introspection was, thereby, made respectable, in that it was correlated with physical states at the neural or muscular level as well as with facial expressions and bodily gestures. Thus was inaugurated a period of psychological research that took seriously questions of emotion, consciousness, and volition. Some, such as Henry Maudsley and W. B. Carpenter, even collapsed the cherished dualism between mind and body, although subsequent investigators, James Sully most notably, were compelled to restore the sacrosanct divide. Evolutionary biology, in the hands of Herbert Spencer, and the German experimental research of Fechner and Weber also played a role in shaping a vibrant community of psychologists in Victorian Britain (see Smith 1973; Jacyna 1981; and Daston 1978, 1982). Although concrete knowledge of neurophysiology then as now was grossly inadequate to the task, the mere presence of appeals to physiology infused the discipline with an aura of scientific objectivity.[9] This in turn served to dissipate some of the religious and ethical debates that surrounded the question of free will.

A contemporary of Mill's, Richard Jennings, also drew a line between the "province of human nature" and the "external world." His *Natural Elements of Political Economy* (1855) is replete with remarks about the nature and scope of the subject, most notably highlighting the importance of psychology in the development of political economy (see White 1994b). In Jennings's view: "All the phenomena of Political-economy are of two kinds, caused severally by the action of matter on man, and of man on matter." Thus: "There occur simultaneously mental phenomena and physical phenomena, mutually connected by laws, to determine which is the chief object of abstract Political-economy." There are, for Jennings, laws of human nature that are as "fixed and invariable" as the laws of nature, and to some extent these have already been discerned by statisticians. But the main point that Jennings drives home is that the phenomena of political economy, such as exchange value, are mental in origin. It is imperative, therefore, to develop psychological inquiry. He even, quite perspicaciously, proposes: "From this law of the variation of sensations consequences will be found to ensue, affecting more or less all the problems of Price and of Production" (Jennings 1855/1969, 21–22, 10, 140, 99–100).

John Elliott Cairnes is often classified as the last prominent economic theorist of the classical era. His *Character and Logical Method of Political Economy* (1875) confronts directly the issue of the epistemological status of the science of wealth. In his view, political economy is on the same par as astronomy. "What Astronomy does for the phenomena of the heavenly bodies," he declared, "Political Economy does for the phenomena of wealth" (Cairnes 1875/1965, 35). Notwithstanding the seeming messiness of the empirical record, everything in the economy is law governed, and the task of the political economist is to discover those laws. With an implicit debt to Darwin's entangled bank, Cairnes asserts that the phenomena of political economy, "the prices of commodities, the rent of land, the rates of wages, profits, and interest, differ in different countries; but here again, not at random. The particular forms which these phenomena assume are no more matters of chance than . . . the fauna or flora which flourish on the surface of those countries are matters of chance" (36).

Interestingly, Cairnes argues that, while political economy draws on both the material and the mental, it is in some sense neither a material nor a mental science. There is an equal dependence on laws from both domains, but, in some unspecified sense, wealth, the true subject matter of political economy, is a domain unto itself: "Neither mental nor physical nature forms the *subject-matter* of the investigations of the political economist. . . . The subject-matter of that science is wealth; and though wealth consists in material objects, it is not wealth in virtue of those objects being material, but in virtue of their possessing value—that is to say, in virtue of their possessing a quality attributed to them by the mind" (Cairnes 1875/1965, 48). This equivocal but clear recognition of the mental dimension of the subject was to be resolved by the early marginalists, notably Jevons, Edgeworth, Wicksteed, and Marshall. With

Jevons, economics was to be placed entirely in the domain of the mental: "The theory presumes to investigate the condition of a mind, and bases upon this investigation the whole of Economics" (Jevons 1871/1957, 14–15). All prices were said to be reducible to the feelings of pleasure and pain at the margin, in terms of "the final degree of utility" (52), as Jevons put it. Moreover, while prices changed as the result of the aggregate effect of individual deliberations, there was no need for a common measuring rod between minds or even for one mind to directly sway another:

> Every mind is thus inscrutable to every other mind, and no common denominator of feeling seems to be possible. But even if we could compare the feelings of different minds, we should not need to do so; for one mind only affects another indirectly. Every event in the outward world is represented in the mind by a corresponding motive, and it is by the balance of these that the will is swayed. But the motive in one mind is weighed only against other motives in the same mind, never against the motives in other minds. . . . Hence the weighing of motives must always be confined to the bosom of the individual. (Jevons 1871/1957, 14)

Jevons has here granted considerable autonomy to individual minds as the source of all economic features of the world.

Even capital was defined in terms of mental attributes, as something fixed in objects through the passage of time and the intentions of the person who uses the object. What is critical is that an object be intended for the production of additional wealth. A loaf of bread can "feed the hardworking navvy, the idle beggar, the well-to-do annuitant" (Jevons 1871/1957, 296). But only in the first case is the bread an object of capital. In sum: "There is nothing which marks off certain commodities as being by nature capital as compared with other commodities which are not capital. The very same bag of flour may have to change its character according to the mental changes of its owner" (285).

Rent and wages were also recast in terms of the utility theory of value, redefined in terms of mental states. Labor was simply the production of utility. It creates nothing material. And rent, as Ricardo had already demonstrated, was essentially a function of the configuration of property relations in a given region. All economic development came from human wants and desires, which are taken to be sui generis; it was in the act of deliberation that economic phenomena were formed and altered.

Francis Ysidro Edgeworth was even more emphatic about the psychological turn taken by economics (see Mirowski 1994a, 5–19). His only book, *Math-*

ematical Psychics (1881), called for a full mathematization of economic theory in terms of the utility calculus and drew direct inspiration from the psycho-physiological work of the German experimental psychologists, such as Hermann von Helmholtz and Gustav Theodore Fechner. For Edgeworth, the first principle of economics is that "every agent is actuated only by self-interest." Pleasure, as the Comte de Buffon had shown, was a property of human evolution and, thus, "an essential attribute of civilisation." Utility was viewed as a kind of energy and, thus, measurable. Humans were deemed "pleasure machines." Indeed, the day could conceivably come when there would be an instrument, "a psychophysical machine, continually registering the height of pleasure experienced by an individual" (Edgeworth 1881/1967, 16, 77, 15, 101). For Edgeworth, economics was firmly rooted in psychology, which was something to relish, not fear. It would bring greater rigor and objectivity to the subject and provide the proper ontological foundation for the newly developed mathematical theory.

Philip Henry Wicksteed came to economics by way of the Unitarian ministry and was so taken with Jevons's new ideas that he hired a private tutor to reacquaint him with the calculus. He also endorsed the psychological turn.[10] In his entry "Political Economy and Psychology" for *Palgrave's Dictionary of Political Economy* (1896), he remarked on the links between the two fields that had recently been forged: "The economist must from first to last realise that he is dealing with psychological phenomena, and must be guided throughout by psychological considerations" (Wicksteed 1910/1933, 767). This is true not only of the analysis of consumption, which gives psychology a "conspicuous place" in economics, but also of all the other areas of the science, such as production, distribution, and money. These latter categories are all governed by the law of supply and demand and, thus, by the psychological questions of satisfaction and motivation. In sum: "The direction taken by economic study in recent years tends to a more express and generous recognition of the close connection between psychology and political economy, and the necessity of constantly keeping in touch with our psychological basis even when pursuing those branches of economic inquiry which appear to be remotest from it" (Wicksteed 1910/1933, 769). It was clear to Wicksteed that the character of economics had changed dramatically since the 1870s and that a central factor in that change was the harnessing of psychological theories.

The final figure to be considered in this overview of Victorian economics is none other than Alfred Marshall (1842–1924), who dominated, not only the last decade of Victoria's reign, but, given his imprint on John Maynard Keynes, the first half of the twentieth century as well. Marshall was very interested in psychology in his formative years and wrote several essays on the subject, essays that remained unpublished in his lifetime. The editor of these papers, Tiziano Raffaelli (1994), has suggested that Marshall took the human mind to be a machine of relative simplicity and, thus, still adhered to some of the basic tenets of the associationist school. Borrowing heavily from Alexander Bain, Marshall believed that most mental connections were grounded in contiguity and similarity. All actions stem from the mind, but, within the mind, there is room for the reassembly of sensations from the external world and possibly some internal machinery as well.

This latter belief may stem from Marshall's affinity for Kant, even though his psychological inquiries were more in keeping with the empiricist tradition of Hume and Bain. With little delving, Marshall tried to draw a line between the mental and the physical: "My psychological facts are independent of my physical facts, although in any hypothesis or theory by which I attempt to connect my psychological facts I shall be indebted at every step to my corresponding physical theories" (quoted in Raffaelli 1994, 113). Deliberation comes because we evaluate and rank future actions differently than present ones. All market phenomena thus come from the mind, and even the formation of capital is essentially the result of an investment of time, of forgoing immediate consumption. A person in an economic context makes very straightforward decisions that revolve around his preference for the present over the future: "Sometimes he is like the children who pick the plums out of their pudding to eat them at once, sometimes like those who put them aside to be eaten last" (Raffaelli 1991, 50). In sum, man produces nothing material, only utility. And it is investment in one's mind that matters most. Physical capital is taken to be subordinate to intellectual capital: "The most important machine is man, and the most important thing produced is thought" (Raffaelli 1991, 52).

This brief tour through the leading Victorian economists has, I hope, lent weight to the significant presence of psychology at the time. Needless to say, there were many other novel developments in the field of economics. Herbert Somerton Foxwell, a contemporary of Edgeworth's, wrote a succinct overview, "The Economic Movement in England" (1887), that also acknowledged the emergence of the historical school (or the discipline of economic history as we now know it), of socialism and Marxism, and of a general shift away from crude laissez-faire reasoning by one and all. I have not discussed these trends here, in part because they have already been addressed by other historians (see Dobb 1973; Kadish 1982; Maloney 1985), and in part because they do nothing to alter the theme highlighted here. Indeed, these three developments reinforce the general transformation of conceptions of the economy as something that need not simply be left to the laws of nature because it can, in fact, be understood and managed. Economic history tended to undercut the belief in laws altogether, by emphasizing the unique and ideographic features of the economic landscape. Socialism was predicated on reform and the refusal to commit the naturalistic fallacy. And the widespread dismissal of laissez-faire principles speaks for itself.

The advent of concentrated appeals to psychology in economic discourse has not been given proper recognition by specialists in the history of economics. The inherent subjectivism of psychology seems to cast a murky shadow over a science as solid and rigorous as economics. Such suspicions date back to the early twentieth century. Irving Fisher, arguably the most prominent analytic economist in the first third of the twentieth century, was strongly opposed to the use of psychological findings in economics (see Chaigneau 1995). But it was the legacy of positivism that nailed the coffin shut on psychology, particularly in the work of Paul Samuelson, who even purged economic discourse of the concept of utility because it was too subjective. Economists have since been content to speak of *revealed preferences* and leave the inner workings of the mind to others (see Mirowski 1989, 222–31, 378–86).

Why was there this fleeting fancy for psychology in Victorian Britain? No similar fancy was found on the Continent in the nineteenth century or in any of the treatises of the seventeenth and eighteenth centuries. There is no one, simple answer. Certainly, Mill's preeminence, and his own familial acquaintance with the subject, played an important role in bringing psychological reasoning to the fore. His enthusiasm for Bain's efforts to join physiology with associationist psychology may have been the critical turning point. Perhaps this was enough to weaken any resistance that might have been posed by Comte and his refusal to have any truck with something as inscrutable as the human mind. Fred Wilson has argued in great detail that Mill assuaged similar doubts and convinced himself that there could be an empirical science of psychology. More important, it was Mill who pointed the way to a purely qualitative analysis of pleasure and, thus, the important notion of an ordinal ranking of inner states (Wilson 1990, 220). Even when later economists such as Jevons repudiated Mill on such grounds as his doctrinal and methodological commitments, they waxed enthusiastic about psychology and upheld the view that one could infer mental states from manifest actions.

One factor that may have sustained this favorable attitude toward the science of the mind in the post-Millian period stemmed from a predilection for the individual as the point of departure for all economic theory. In a nutshell, the early neoclassical economists dissolved economic classes as the unit of analysis and built their model of the world from individuals. Whereas with Ricardo the central question was how to divide the pie between the landlords, the capitalists, and the laborers, with Jevons and Marshall the key question became one of maximizing individual utility and taking the aggregate to arrive at meaningful claims about social welfare. There were no more classes; indeed, it was proposed that every laborer might also be an owner of capital and certainly partook in the cash and credit nexus. To put it another way, individuals rarely made an appearance on the stage of the classical economists; the forces of capital accumulation and the ongoing rise of commerce dwarfed individual differences. For the early neoclassical economists, individuals—albeit faceless and nameless individuals—were the principal source of all economic phenomena, which, as we have seen, were mental, not material. Moreover, the properties and motions of market phenomena were directly the result of the fact that human minds differed one from the other, at least in terms of their evaluations of pleasure, pain, time, and risk. Introspection and studies of the mechanisms of the mind were, thus, just the license required by Victorian economists to reorient the discipline around individual agency.

Enthusiasm for psychology and the inner life of the mind was also to be found among natural scientists of the time, although the form taken seemed to run parallel to that found among economists rather than impinging directly on it. A striking feature of the community of physicists in the latter half of the Victorian period was their reverence for the world beneath the given of experience. Perhaps in reaction to Comtian positivism and its atheistic associations, Victorian physicists delighted in the spiritual dimensions of the *unseen universe*, as it was dubbed in 1875 by Balfour Stewart and Peter Guthrie Tait. With the formulation of the law of the conservation of energy, the luminiferous ether became the seat of the electromagnetic field, of light, and of heat—and, most important of all, of the deity (Heimann 1972; Wilson 1977; Wynne 1979). This was the period when British physicists such as Oliver Lodge and George Gabriel Stokes (and economists like Edgeworth) were smitten with spiritualism and psychics. Such predilections for the mental resonated well with the declarations of the psychologists of the time, notably W. B. Carpenter and Fechner. Mental phenomena were readily joined to energy and its various manifestations.[11]

In late-Victorian political economy, as we have seen, individualistic psychology was quite pronounced. This was in sharp contrast to political economy before 1830, where reason was subordinate to the passions. In the classical theory, the individual mind did not make choices that determined the pricing and distribution of economic goods. These were, rather, the result of the configuration of large groups of people, distinguished by their need to labor in rhythm with the cycle of the harvest. Moreover, Hume and Smith took humans to be much more like animals, suggesting in numerous passages that human intelligence was merely refined animal instinct (see Pitson 1993; and Schabas 1994a). As Hume argued persuasively, our basic belief in causation is formed by nothing more than custom, which is part of the same continuum as animal instinct. Habit and custom, not reason, are what propel all our inductive knowledge of the world:

> It is impossible, that this inference of the animal can be founded on any process of argument or reasoning, by which he concludes, that like events must follow like objects, and that the course of nature will always be regular in its operations. For if there be in reality any arguments of this nature, they surely lie too abstruse for the observation of such imperfect understandings; since it may well employ the utmost care and attention of a philosophic genius to discover and observe them. Animals, therefore, are not guided in these inferences by reasoning: Neither are children: Neither are the generality of mankind, in their ordinary actions and conclusions. (Hume 1748/2000, 80)

Not only has Hume reduced most human thought to the level of that of sheep, but he has also suggested that humans are like sheep in another way: we are all more or less alike in terms of our mental faculties. By nature, we differ in no significant way one from the other. It is education and happenstance that make the difference, as Smith remarked: "The difference of natural talents in different men is, in reality, much less than we are aware of. . . . The difference between the most dissimilar characters, between a philosopher and a common street porter, for example seems to arise not so much from nature, as from habit, custom, and education" (Smith 1776/1976, 28–29). Eighteenth-century approaches were much more inclined to treat human reason as derivative of more fundamental natural instincts and propensities than were Victorian, which granted so much efficacy and power to the mind. This had important implications for concepts of economic phenomena. In the Enlightenment, they were much more directly linked to natural processes, while, in the latter half of the nineteenth century, they were viewed as the product of human deliberation.

CHAPTER EIGHT

Denaturalizing the Economic Order

Physics and economics . . . are both disciplines with imperialist tendencies: they repeatedly aspire to account for almost everything, the first in the natural world, the second in the social.

—Nancy Cartwright, *The Dappled World*

EVOLUTIONARY THEMES

Collingwood's (1945, 10) assertion that the nineteenth-century view of nature was essentially evolutionary still rings true, although the reasons that he adduced have become less compelling. Collingwood identified Herbert Spencer and Henri Bergson as the leading proponents of this evolutionary trend, and he saw them as progeny of Hegel. As a result, he also emphasized the revival of ideas of purpose and telos in concepts of nature and the decline of mechanistic modes of thinking. Hardly anyone reads Spencer or Bergson now or thinks of Hegel as central to concepts of nature, but, on this last point, Collingwood was sound. Certainly by the latter half of the nineteenth century, thermodynamics had taken physicists quite a distance from the mechanical worldview (see Harman 1982, chap. 6; and Buchwald 1985). Moreover, for most nineteenth-century philosophers of nature, purpose was paramount. What made Darwin stand out among his contemporaries was his relative immunity to teleological thinking.

It would seem reasonable to assume that political economy too became evolutionary in orientation. Certainly when economists think of joining their subject with evolutionary biology, it is Alfred Marshall who springs to mind for his celebrated remark: "The Mecca of the economist is economic biology rather than economic dynamics" (Pigou 1925, 318). His endorsement of the same motto as Darwin regarding nature's inability to take leaps has also been taken to suggest that Marshall was profoundly influenced by Darwin (see Niman 1991). It is certainly very tempting, and quite easy, to tell the following story. Darwin's *Origin of Species* (1859) made of biology such a respectable scientific field that economists could turn to it, instead of physics, as a model. More specifically, Darwin's insights greatly reinforced economists' long-standing appeals to competition, equilibrating mechanisms, and historical explanation. Most of all, his thoroughgoing materialism transformed our conception of human psychology and morality. Both were products of our evolutionary history and, thus, at bottom, just refined instincts. Economists could discard, once and for all, appeals to a human nature designed by a deity.

There is little evidence, however, that Darwinian biology shaped the content or even the broader context of early neoclassical economics, particularly as represented by Marshall. In asserting this disjunction, I do not wish to suggest that economic theory and the theory of evolution have nothing in common. On the contrary, as we saw in the preceding chapters, biological reasoning and economic reasoning have been closely conjoined since the Enlightenment. But there is probably more Linnaeus in Adam Smith than Darwin in Marshall. Even John Stuart Mill's conception of the economic order is arguably more at one with the broader tenets of Darwinian biology than Marshall's—despite having been forged without knowledge of Darwin.[1]

It is important to draw a distinction between a favorable regard for Darwin, which was the case for most of the prominent British economists alive in the decades after 1859, and a genuine absorption of Darwinian mechanisms into economic theory. Mill remarked in a letter of 1860 that Darwin's book "far surpasses my expectation" (Mill 1972, 695), and Jevons compared Darwin (and Spencer) to Newton in terms of "revolutionising . . . all our views of the origin of bodily, mental, moral, and social phenomena" (Jevons 1877, 762). Marshall first read Darwin—with much enthusiasm—during his "apprenticeship years" as a member of the Grote Club (Whitaker 1977, 194), and, in the opening sections of his *Principles of Economics,* he explicitly acknowledges the importance of Darwin's theme of historical contingency (Marshall 1890/1920, 42).

Camille Limoges and Claude Ménard have argued that Darwin's (and Ernst Haeckel's) insights on the division of labor permeate Marshall's insights on industrial organization in book 4 of the *Principles* (see Limoges and Ménard

1994). But recall that these ideas came to Darwin from Henri Milne-Edwards, who in turn gleaned their importance from Say. For this and other reasons, Limoges and Ménard's argument that Marshall's economics is explicitly Darwinian is not convincing. In the Marshall Papers, there is an early draft of the section of the *Principles* dealing with the subject of the division of labor that makes not one single reference to biology.[2] Not that one would have expected any such reference: the differences between Darwin's notion of the division of labor and Marshall's are pronounced. First, Darwin's division of labor was random, with no end in mind except a viable organism. In that respect, Darwin had already broken away from the Smithian model. But so had many of Marshall's predecessors, such as Ricardo and Mill. There is no evidence that Marshall grasped this from Darwin directly. Second, there is no clear analogue to the Darwinian mechanism of natural selection in Marshall. Marshall could point to the diversity of firms, but what filled the place of the laws of heredity, let alone the principle of superfecundity? Third, as Limoges and Ménard acknowledge, Marshall's concept of the representative firm was purportedly a reversion to essentialist thinking, despite Darwin's own brilliant predilection toward populationist processes.

Darwin was a palatable tonic for economists if only because his analysis of the economy of nature read, as we have seen, like classical economics applied to the natural realm. But this was a shortcoming as well as a strength. Precisely because Darwin's notion of the struggle for existence in the economy of nature resonated with extant economic doctrine, economists may have been disinclined to take the trouble to understand the intricacies of his theory. That seems to be the case with both Mill and Marshall. Darwin's theory of descent with modification is, by any standards, a very sophisticated piece of reasoning. Arguably, it takes years to absorb the details of the theory—for example, the concept of fitness—and to appreciate its explanatory richness.[3] Few, if any, at the time absorbed Darwin's populationist notions of a species. And some, such as Jevons, found his reluctance to impute progress to the evolutionary scheme quite troubling (Jevons 1890, 273–74). Spencer's sanguine gloss on the biological process was much more appealing to Victorian economists, who were the first to insist that they pursued their subject as the royal road to social improvement.

On the few occasions when Marshall draws analogies to things biological, most of his images, such as the cycles of birth and death, might as well have come from Aristotle. In the unpublished "Law of Parcimony," Marshall harks back to Aristotle's maxim that nature does nothing in vain (see Marshall 1867). Moreover, the essay "Mechanical and Biological Analogies in Economics"

(Marshall 1898) contains not one single reference to evolutionary biology. And, apart from the analysis in book 4 of the *Principles* and Marshall's appreciation of historical contingency, there is but one reference (Marshall 1890/1920, 495) to the principle of the survival of the fittest, a phrase that originated with Spencer, not Darwin. A. L. Levine's careful exegesis of Marshall's biological fragments shows only a cursory appreciation for Darwinian processes (Levine 1983). The discussion of a social organism and appeals to statics and dynamics more likely came from Auguste Comte, whom Marshall much admired.[4] Certainly, Marshall's numerous claims that appeals to biological analogies ought to be made only after economics has reached a certain level of maturity, via appeals to physical analogies, has a distinct Comtian ring to it.

Marshall makes much ado about human wants and activities in the opening chapters of the *Principles*. As John Dennis Chasse has argued, these passages are emblematic of the broader philosophical framework of Marshall's work. Like Marx, Marshall was struggling to come to terms with our species being, and, again like Marx, the debt is to Hegel far more than to Darwin. Benjamin Jowett, the preeminent Plato scholar of his day and one of the few who held intellectual sway over Marshall, expressed a distinct pleasure in seeing "a considerable element of Hegelianism" in the *Principles* when writing Mary Paley Marshall in 1890 (see Whitaker 1996, 337). Scholars have since substantiated Marshall's appreciation, not only for Hegel, but also for Spencer (see Pigou 1925, 11; Whitaker 1977, 193; and Groenewegen 1990).

How different was Spencer from Darwin? According to Robert Richards, quite a lot (Richards 1987, 291–94). Spencer had first embraced evolutionary ideas in the early 1850s and extended them well beyond the biological realm. Moreover, he was a thoroughgoing Lamarckian. While Darwin also accepted the principle of the inheritance of acquired characteristics, Spencer was far more reluctant to assimilate the principle of natural selection because of its purposeless and random implications. In other words, he was never a Darwinian, even granting Darwin's Lamarckian leanings. Interestingly, Jevons usually lumped Spencer and Darwin together when referring to evolutionary biology, but, in his discussion of human nature, it is Spencer who receives a full endorsement, Jevons attributing the recent "revolution" in moral philosophy to him (Jevons 1890, 289).

Thanks to the perceptive work of Robert M. Young, historians have also come to see Darwin and his contemporaries as more than a solution to the problems of extinction and the geographic distribution of life-forms. As Young has put it, the central debate of the nineteenth century—thrashed out in many disciplines, including theology, political economy, anthropology, and psychology as well as biology—was "man's place in nature." More important, disciplinary boundaries were, at the time, highly permeable. As Young has put it: "In the nineteenth-century debate there was an intimate mixture of psychological, social-philosophical, biological and theological issues" (Young 1985b, 78). Certainly, the wall between the natural and the social sciences was far less firm than it is today.

Perhaps we have been looking in the wrong place. Perhaps it is the broader implications of Darwin, rather than the specific mechanism of natural selection, that might have influenced neoclassical economics. With Darwin, we have for the first time a detailed argument that man is related by descent to every other living form and, thus, that even human intelligence and morality are simply refined instincts (see Durant 1985). Are there signs that this very profound and novel perspective made its way into economics?

I think not, at least in the early neoclassical era of Marshall. As we have already seen, comparisons between human and animal behavior reach back to at least the eighteenth century. Think of Bernard Mandeville's *Fable of the Bees* or the writings of David Hume. It was never (or rarely) imagined, however, that human nature might resemble animal nature because of a common ancestor. Comparisons were normally drawn in order to emphasize the uniqueness of human reason and the universality of other propensities. In the opening sections of *The Wealth of Nations,* Adam Smith notes that the faculties of reason and speech are to be found only in humans, hence our ability to engage in economic exchange. I hesitate to claim that the converse of this is not to be found in the neoclassical literature. Indeed, Jevons once quipped: "I should not despair of tracing the action of the postulates of political economy among some of the more intelligent classes of animals" (Jevons 1876/1905, 197). But we can gather from a letter to his wife that Jevons intended this more as a joke (see Jevons 1977, 182). No one seems to have developed the idea at the time.

More modestly, Darwin's theory implied that one should seek the roots of human nature in human physiology. This message may have been what propelled Jevons and Edgeworth to adopt a reductionist view of psychology, although I know of no concrete evidence to confirm this connection. Jevons worked much more along the lines of Benthamite introspection, and Edgeworth drew inspiration from the German school of experimental psychology. In fact, an evolutionary psychology does not necessarily entail the reduction of mind to matter. According to Young, the case of David Hartley and Erasmus Darwin demonstrates that "associationist psychology, suitably extrapolated, becomes evolution" (Young 1985b, 71). In short, evolutionary biology underdetermines the theory of psychology that one ultimately endorses. A commitment to viewing human traits as refined instincts still permits one to stake out many different positions along the material–mental continuum.

Robert Richards's excellent study of the subject gives the impression that Darwin was forced to resort to a considerable amount of hand-waving in this as in other areas of his work. His conviction that our social and sympathetic capacities were instinctive rather than learned was supported by observations of other animals. What enabled us to become moral creatures was our ability to deliberate. If another species were to develop a similar ability to reason and couple that ability with social instincts, then it too would acquire a moral sense (Richards 1987, 210).

In this emphasis on deliberation, it would seem that Darwin was at one with Bentham. But, as Richards has persuasively argued, Darwin's theory "overturned utilitarianism" (Richards 1987, 218). The differences much outweighed the similarities, as Darwin himself fully realized. The good was to be grounded, not in the consequences of actions, or in a self-interested view of human action, but in what nature deemed viable for the community as a whole. Nor was pleasure the mainspring of human action for Darwin. Individual pleasure has no cash value in the evolutionary scheme of things (Richards 1987, 217–42).

An interesting, and relevant, discussion of Darwin is to be found in Marshall's "Law of Parcimony," which in its twenty-five pages contains more musing on Darwin than any of Marshall's other writings.[5] The law of parsimony itself—put forward in 1837 by Sir William Hamilton (who employed the more common spelling)—dictated that no more causes or forces should be assumed than are necessary to account for the facts. For reasons that are obscure, Marshall explores it in the context of Darwinian biology and the psychology of Condillac and Bain, among others. The thrust of this early discussion is to place severe limitations on the reliability of both areas of inquiry. Marshall draws attention, for example, to Darwin's "naive simplicity" and his tendency to "exceed the authority which experience can give" his fundamental principles (Raffaelli 1994, 97). Darwin can only conjecture as to what might have been the common ancestor of two living species, and, when it comes to complex organisms like the human eye, he runs into severe difficulties (as Darwin himself noted).

For our purposes here, the most interesting portion of "The Law of Parcimony" is the frequent comparisons and contrasts drawn in it between biology and psychology, particularly on the subject of methodology. Marshall recognizes the importance of analogical reasoning, coupled with Occam's razor. There is, in his view, "a remarkable analogy and a still more remarkable difference between the fundamental methods of his [Darwin's] investigation and those of psychology." Whereas, in Darwin's case, the phenomena are homogeneous, when it comes to psychology, Marshall submits, the phenomena are irreducibly heterogeneous. Psychological inquiry thus runs up against major obstacles: "Between the idea of a sensation and an idea of similarity between sensations there is no relation" (Raffaelli 1994, 96, 99). And, when it comes to connecting the ideas formed by the different perceptual faculties (like taste and smell), analogical reasoning breaks down altogether.[6]

Comte had already voiced, in considerable detail, the difficulties of connecting social physics to physiology, although he maintained that the "homogeneity" of the phenomena offered a glimmer of hope (Lenzer 1975, 95). Social inquiry must start with the individual, but social physics proper has its own set of phenomena, those that pertain to the social organism. Post facto, it is difficult to map Darwinian evolution and associationist psychology onto the Comtian ladder, although they would, presumably, fall somewhere in the nexus of social physics and physiology (with the proviso that they met positivist standards in the first place). Perhaps it was this task that prompted Marshall's train of thought, although Comte is not explicitly cited. Spencer is mentioned a couple of times, and he too dwelt on the problem of the homogeneity of phenomena for a given domain of inquiry. In any event, Marshall seems to have shut the door quite firmly on linking any further those two disparate branches of knowledge. If Darwin left his mark on Marshall's economics, it was not via psychology. The negative tenor of "The Law of Parcimony," together with Darwin's expressed opposition to utilitarianism, puts the burden of proof on those who would wish to urge a strong influence of Darwin on Marshall's conception of mental processes and moral principles.

Another major consequence of Darwin's theory that commentators have noted is that it lent enormous force to a belief in the uniformity of nature. As Darwin himself noted: "When we look at the plants and bushes clothing an entangled bank, we are tempted to attribute their proportional numbers and kinds to what we call chance. But how false a view is this!" (Darwin 1859/1968, 125). Every speck of life on that bank is there because of the laws that govern the organic realm.

Mill, however, had already advanced this doctrine of the uniformity of nature in the *System of Logic* (1843), some fifteen years before Darwin published the *Origin of Species* (1859). Taking stock of the long list of extant laws in the natural sciences, Mill contended that there could be no other explanation of this fortunate state of affairs than the fact that nature was, indeed, uniform. Jevons advanced much the same argument, pointing to the established body of laws in physics. It does not appear to have been the case, then, that Darwin's findings were critical in elevating convictions among economists on this matter, although they were certainly welcome reinforcement, as we saw in the case of Cairnes.

Peter Bowler has argued that Darwin's specific version of evolution was never predominant in the Victorian period. Since the so-called Darwinian synthesis of the 1930s and 1940s, we have come to pay homage to Darwin above all (although we have also had to ignore his Lamarckian leanings and his allegiance to continuous variation), but, during the mid-nineteenth century, there were many other "transformationist" notions circulating among naturalists. Lamarck's ideas were taken seriously, as were the ideas of Geoffroy Saint-Hilaire and Richard Owen. And there were others who adhered either to the transformationism of *Naturphilosophie* or to outright creationism. Bowler has put it as follows: "If Darwin's radical insights catalyzed the transition to evolutionism but were ignored by most 'post-Darwinian' thinkers, are we justified in treating the emergence of the selection theory as the key event in the theory's history? . . . I suspect that the world is not yet ready for a survey of evolutionism in which Darwin does not play a pivotal role. Nevertheless, in my view, our current fascination with Darwin's discovery of natural selection is at least in part an artifact of modern biology's commitment to the synthesis of selectionism and genetics" (Bowler 1989, 24).

Even if this is an overstatement, it most surely holds for Darwin's views on human nature. Arguably, not one naturalist at the time agreed with his specific views on human psychology and morality (Richards 1987, 234). Nor were any willing to accept his repudiation of purpose in nature. The task that I set myself here seems to have been to search for something that is nonexistent. Darwin could not have influenced Mill or Marshall because he was not influential period. I do not wish to go that far, however. Bowler's position, in my view, is the product of malaise brought on by oversaturation. Of course Darwin was a central figure in Victorian England, as Mill, Jevons, and Marshall fully recognized. My claim is only that it is difficult to identify the specific Darwinian elements influencing economic thought at the time. The reason is simple, of course. Lyell and Darwin were already suffused with imagery from classical economics. There were virtually no more gains from trade to be had.

Scott Gordon once noted that, while Marshall voiced the prospect of leading economic theory toward the true Mecca of biology, no one has yet managed to carry out the task (Gordon 1973, 248). More recently, A. W. Coats has echoed this sentiment (Coats 1990, 170). But, in fact, that place was reached long before Marshall, in the mid-eighteenth century, and was then shrouded in clouds. Marshall's appeal may be better understood as one of his Romantic sighs recalling a time now lost, in part, I conjecture, because of his Hegelian conviction that the fundamental laws of the economic realm are historical.

Indeed, were it not for those remarks made by Marshall, we would probably not have bothered to look for Darwin's influence in the first place. Certainly, few, if any, of Marshall's contemporaries or immediate successors adopted his appeals to biology or even embraced his predilection for historical laws. If the current trend in the history of biology is closer to the truth, we can safely abandon this quest since Darwin was not supremely important after all, at least not until the 1930s and 1940s. And, by then, neoclassical economics had matured, both as a body of knowledge and as a professional unit, such that its external membranes were considerably more impervious to infection from such distant fields as evolutionary biology.

Classical economists took the economy to be a natural entity and saw *homo economicus* as a creature of animal passions and instincts bent on outcomes such as excess population and the dreaded stationary state that were at odds with the dictates of reason. Subsequent economists, such as Mill and the early neoclassicists, took man out of nature. The economy was seen to be the result of rational agency and, thus, no longer directly governed by natural forces. Economic well-being was not like the ebb and flow of the oceans, as Hume had once suggested (see Rotwein 1970, lix), but something that could be planned, if not controlled. Humans need no longer struggle against nature. Recall Mill's paean: "The ways of Nature are to be conquered, not obeyed" (Mill 1874/1969, 380–81). Victorian economists thus tugged in a different direction from Darwin and the social Darwinists. From our own vantage point more than a century later, the formal similarities in game theory notwithstanding, they still appear to be doing so.

STABILIZING THE ECONOMY

One of the most striking features of post-Marshallian economics is its aversion to historical modes of thinking (only the American school of institutionalism is an exception). Whereas the eighteenth-century economists, Hume and Smith most notably, established many of their principles through the use of historical analysis and believed them to be developmental, the neoclassicists believed their principles to be timeless and universal in application. Humans have always been rational utility maximizers, and, hence, they have never evolved. The essential phenomena of an economy have been present in every known civilization. Even the field of economic history, or *cliometrics,* as it has come to be known since the 1960s, has in many respects refurbished itself as a testing ground for the fundamental principles of neoclassical economics rather than cultivate genuine historical insight (see Schabas 1995b). This is all the more true of the new institutionalist school, where most of the historical record is distilled to the forward march of decreasing transaction costs (see Rutherford 1996). Institutions are merely the sum of rules, and the rules themselves tend to reflect a timeless rationality. Compared to so many other areas of intellectual inquiry—anthropology, psychology, philosophy—economic thought since the 1870s shows little evidence that it has absorbed evolutionary thinking. Quite the opposite, economists have embraced the view that bygones are always bygones, that the economy is always re-creating itself and, hence, that the past does not matter.

This claim may seem surprising given the sizable analogical trade between biology and economics. Jack Hirschleifer's seminal "Economics from a Biological Viewpoint" emphasizes the strong similarity between the "fundamental organizing concepts of the dominant analytical structures employed in economics and in sociobiology" (Hirschleifer 1977, 1–2). But most of what Hirschleifer identifies harks back to the concept of the oeconomy of nature, with the exception, perhaps, of optimization. Partly for this reason, heterodox economists such as Kenneth Boulding (1981) and Geoffrey Hodgson (1993) continue to demand that economics embrace an evolutionary standpoint, by which they mean a nonequilibrium approach that captures developmental processes. Economic theory may not have had a strongly evolutionary content since Thorstein Veblen's *Theory of the Leisure Class* (1899).

Neoclassical economics has also acquired a strong apolitical character to match its divorce from history. Marshall chose the title *Principles of Economics* carefully. The discipline was to become known as *economics,* not *political economy.* It was no longer overtly tied to political imperatives. When Marshall helped found the British Economic Association in 1891 (later renamed, somewhat ironically, the Royal Economic Association), he took great pains to ensure that the charter membership spanned the political spectrum and, thus, tolerated political pluralism (Coats 1968). Unlike their counterparts in the United States, who were deeply divided by political allegiances, British economists had matured past the point of political dogma (Haskell 1977). At least that was the impression that Marshall sought to convey. Certainly, he put to rest any remaining controversies that Ricardo and the Ricardian socialists had stirred up in the 1820s and 1830s. As John Maloney has argued, Marshall coated economic theory and its professional trappings with a veneer of ideological neutrality, and he did such an excellent job that little to no further justification has ever been required (Maloney 1985).

With the early neoclassical economists—Jevons, Edgeworth, and Marshall—the economy was placed squarely within the realm of human agency and institutions. There were still some physical features to economic worlds, but human deliberations became the proximate causes of all economic phenomena. Any given price (for there is no longer a natural price) is essentially the result of hedonic balancing acts rather than of labor inputs. Even labor and capital are recast in terms of utility and, thus, assessed subjectively. Production takes second place to consumption, and services are put on the same plane as commodities. Gone are Mill's distinctions between productive and unproductive labor. Economics has become a science of rational choice.

Contemporary consumer choice theory is strongly wedded to the metaphysical doctrine of free will. Individual preferences are sui generis; there is no causal account of their formation, let alone transformation. Human agency is, thus, grounded in genuine choice between alternative goods and services. Even preferences for time and risk are fully enshrined in the minds of individuals (see Hausman 1992; and Davis 2003). Part and parcel of this position is a commitment to methodological individualism or reductionism more generally. Neoclassical economists maintain that all aggregate phenomena can, in principle, be reduced without remainder to the activities of individuals. The unemployment rate is nothing more than the sum of those individuals who declare themselves unemployed. Emergent phenomena, or social facts of a Durkheimian sort, are not part of the picture.[7]

Economic texts of the eighteenth century, by contrast, downplayed the role of the individual. Insofar as economic phenomena—money and markets—were joined with physical nature, they were governed by laws that operated at a more holistic level. It was usually as a member of the group—say merchants or farmers—that economic relations transpired. Indeed, the material conditions of economic modes of production, such as hunting or farming, were strongly determinative of individual traits, as evident in, for example, the four-stages theory of the Scottish moral philosophers. For Mandeville, Hume, and Smith, individual efforts to scheme and influence the course of history were repeatedly trumped by social forces and practices.[8] Hume went even further and denied the existence of free will. Social conditions, nationality, gender, and economic station are what most determine our character, and, as a result, character types are remarkably robust over time and space: "The knowledge of these characters is founded on the observation of an uniformity in the actions, that flow from them; and this uniformity forms the very essence of necessity" (Hume 1739–40/2000, 259). As a result, efforts to change our character prove mostly fruitless, and what character change does take place is due mostly to factors beyond our control (see Russell 1995, chap. 9). We are much more governed and determined by the external world than by any traits specific or unique to the individual.

For the eighteenth-century economists, economic regularity stemmed, not from the uniformity of individual reason, but from the cohesive nature of human groupings in conjunction with nature. This is most striking in Quesnay's *tableau économique,* where farmers, artisans, and landowners unreflexively perform their predetermined roles. Methodological holism is also pervasive in the work of Hume and Smith. Think of Smith's masters, who are "always and everywhere in a sort of tacit, but constant and uniform combination, not to raise the wages of labour" (Smith 1776/1976, 84). Humans were part of the Linnaean oeconomy of nature. There was no sharp distinction between physical nature and economic activity.

Needless to say, order and structure existed for most eighteenth-century economists because of the deity. God had created everything, not just the earth and its creatures, but human nature too. For the Physiocrats, the circular flow of the *tableau* was part of the natural order. For Smith, there was a natural progression to the accumulation of wealth in a given region that followed our refinement from cultivator to manufacturer. With the advent of a more secular age, heralded in part by Hume and in full by Ricardo and Mill, most economists no longer looked to economic phenomena as the direct product of the deity. For early nineteenth-century thinkers, *the economy* was still governed by physical nature and subject to a providentialist sensibility. Only over time did human reason and agency supersede divine intention.

To a significant degree, late-Victorian economists repositioned their concept of the economy. The economy was depicted in terms of social institutions, man-made through and through. To put it most emphatically, the economy went from being a natural entity to being a social one. This did nothing, however, to diminish the high esteem for and confidence in the scientific standing of political economy among its practitioners. If anything, it suggested that, by discovering the laws that governed the production and distribution of wealth, economic theorists might also be in a position to change social arrangements. One of the apparent gains of denaturalizing the economic order is that the economy is no longer something beyond our control. Our conviction in the existence of an economic order is no longer predicated on laws of the natural realm.

Assuming that my account is correct, that eighteenth-century theorists treated economic phenomena as much more closely aligned with physical nature than did those of the late nineteenth century, what would this signify? The status of the natural sciences and the virtuous traits associated with nature tend to bring nothing but honor to this earlier epoch, or so one might infer. I would not wish to convey this impression, however. The fact that economic theory underwent a denaturalization over the course of about a century carries with it no deeper meaning. It was a purely contingent process, fueled by relatively independent developments in the natural sciences, by an increasing secularization of European thought, and by the actual development of the economy. Discourses appear to benefit from analogical reasoning, but the channels of trade seem highly accidental (see Cartwright 1999, 1; and Lagueux 1999).

It may even be impossible to sort this out precisely because the phenomena of economics are laden with metaphoric baggage, both natural and social. That there are robust and enduring phenomena that are primarily economic and, thus, deserving of a distinct inquiry called *economics* (or *political economy*) appears to me as unassailable an assertion as can be found in the history of science. Wealth, money, trade, and taxes are some examples of phenomena that have been central to the discourse for centuries, if not millennia; population growth, slavery, market forces, and profit rates are more short-lived, relatively speaking. I am taking Ian Hacking's sense of a *phenomenon* as opposed to a *datum* (see Hacking 1983). Every price in the market is another entry in a data set, but price itself is a phenomenon, an enduring category that lies at the core of economic inquiry. One might think of such phenomena as price as receptacles that acquire different characteristics and, thus, in that sense may well go from being positioned in physical nature to being positioned squarely in the human realm.

Hacking draws a distinction between natural and artificial phenomena, defining the line of demarcation in terms of the presence of human agency (Hacking 1991). But so much of his work emphasizes the role of laboratory apparatus in the creation of scientific phenomena. In physics, for example, "the Hall Effect does not exist outside of certain kinds of apparatus" (Hacking 1983, 226). Hacking argues that this inextricable link with laboratory apparatus has become increasingly the case with modern physics. This, then, implies that Hacking's own distinction between natural and artificial phenomena would dissolve under scrutiny.[9] There is no difference between the Hall effect and the Wicksell effect. In both cases, the human hand is all too visible in creating and measuring the phenomenon in question. Nevertheless, for most of the economists covered by this book, the natural/artificial distinction was believed to be firm. Whereas classical economists viewed their phenomena as belonging to an external physical nature, neoclassical economists believed their discourse to be composed of artificial phenomena. This is evident in the classical concept of a natural price, for example. A natural price was determined by natural costs, which were determined by the cycles of the harvest and population, which were determined by the climate and sexual passions and natural resources. Price theory today steers a different course, emphasizing the role of demand and utility and the deliberations of a firm toward the end of maximizing profits.

If there is any single enduring phenomenon that lies at the core of economic discourse, it is the interest rate, which has been manifest since antiquity, notwithstanding various taboos against usury. For some, of course, this is an odd way to view matters since the interest rate is a man-made phenomenon, unlike, say, the orbits of the planets. Yet the interest rate is also a number and, in that respect, is less contingent on measuring apparatus or political rule than is, say, money. Moreover, the real interest rate has strong natural associations, linked to population growth, estimations of time, and the accumulation of physical capital, which ultimately enhances the well-being of our species. All three of these associations have extensive links with physical nature and the discourses of biology and physics that attempt to make sense of it. It is not so outlandish, then, to conceive of the interest rate as contiguous with physical nature and, thus, treat it in the same terms as the natural sciences. In recent decades, ecologists have adopted game-theoretic models that do just this, namely, impute an interest rate to foraging and gathering processes as a form of capital accumulation (see Stephens and Krebs 1986). Similarly, money, while it may be a human invention and, thus, artificial, has attained "a reality as unyielding to an individual's will as any natural phenomenon" (Foley 1987, 248). Insofar as money is emergent, a force at the macroeconomic level, it tends to pursue an independent track, eluding most efforts at institutional control, or so Charles Goodhart has argued (see Goodhart 1984). My hunch is that the "natural" component of these phenomena is undergoing a revival of late, but that must await further study.

One critical factor in the production of economic phenomena is the role of human agency. As long as differences are diluted as individuals are absorbed into larger groups and deliberation is downplayed, in short, as long as human agency is not the proximate cause of these phenomena, the propensity among theorists is to plant economic discourse in the setting of physical nature. Again the picture provided by the Physiocrats comes to mind. The circulation and growth process depicted by Quesnay's *tableau* does not have room for individual deliberation. Each sector—farmers, landowners, and artisans—carries out its predetermined role with the regularity of a clock (see Charles 2004). All are drawn into the process by the allure of nature's gifts, which allow us to reap two seeds at harvest for every one sown in the spring. If the mercantilists were, in fact, prone to zero-sum thinking, it was appeals to nature that subsequently broke that conceptual barrier and enabled Enlightenment political economists to discern genuine economic growth (Brewer 1995). The account offered here might be reduced to little more than a story of prolonged indebtedness. It took time for nature's gifts to be fully appreciated and woven into the economic literature, hence the conceptual trajectory of the notion of nature's relative role in manufacturing from Smith to Ricardo to Mill. The very fact that Ricardo harped on about the fact that nature's gifts are made "generously and gratuitously" is telling (Ricardo 1817/1951, 76n). That economists came to discount nature's beneficence and ceased to acknowledge any indebtedness on that score came to pass in the second half of the nineteenth century, but it is important to bear in mind that the process transpired over about a hundred years. Now, whatever accounts one might have of nature, ecological or otherwise, a central verity is human control and domination (see Soper 1995). Nature, as Latour and others have argued, is inherently political, but it is a political instrument at the mercy of human reason and agency, or so our economists would have it.

Economic theory continued to borrow heavily, both conceptually and methodologically, from the natural sciences well into the first part of the twentieth century. As Philip Mirowski documented in his celebrated *More Heat Than Light* (1989), there were numerous cases of what he calls *daylight robbery* of scientific material by economists such as Edgeworth, Walras, or Fisher. Mirowski's more recent *Machine Dreams* (2002) has documented the extent to which postwar economics was linked to cybernetics, and in this sense too it is fair to say that economic theory still draws inspiration from the natural or formal sciences. There are some important differences, however, from two or three centuries back. While there is ongoing analogical and methodological trade between economics and the natural sciences, the phenomena themselves are cordoned off into separate spheres. Economic phenomena are the product of human deliberation and human institutions, which are in turn largely independent of evolutionary conditions. If an ecologist adopts a model from economics, there is little to no indication that the causes of the phenomena under investigation are the same. Nor does economics intersect with the physical sciences at the ontological level. Wealth has become a purely mental phenomenon—the pursuit of maximal utility—and has nothing to do with the laws of physics, as Mill once conjectured.

Mirowski has emphasized that the resemblance between neoclassical economic theory and physics is superficial (Mirowski 1989). Insofar as economists have adopted the methods and concepts of physics, they have remained at the level of formal representations. In short, there is little to no depth to the analogies; they are pure simulation. When physicists learn about the Fisher equation, which attempts to mimic the Boyle-Charles law in thermodynamics, they are struck by the misuse of the term *velocity*. Money may circulate in the economy, but it does so in a highly discontinuous fashion. The widespread use of Lagrangian and Hamiltonian techniques in economics also begs numerous questions about the legitimacy of claims about the divisibility and continuity of the physical objects and mental activities in question (see Hausman 1992). As Mirowski has shown, the mathematical methods make sense only if utility is a kind of potential energy, but no one has managed to go beyond the metaphoric veneer in making this assertion (see Mirowski 1987, 90). A possible path toward salvaging this situation might lie in an appreciation of the basic dimensions of space and time, which hark back to Jevons (see Reid 1972). It may well be that capital, utility, even money, can, thus, be reduced to physical constituents with spatiotemporal coordinates. Contemporary economists have not pursued this line of reasoning, however, and perhaps for good reason.

My thesis of denaturalization maintains that the economic order as articulated by leading theorists has moved far away from concepts of physical nature. In other words, whatever gives order to the economy in the first place, whatever gives rise to the central phenomena that warrant an inquiry known as *economics* (or *political economy* in former centuries), is no longer closely linked to physical nature. The economy as it is conceived by economic theorists is remarkably detached from the physical world as it is conceived by practitioners of the physical and life sciences. The two spheres hardly intersect at all, and the very manner in which order is motivated in them stems from different and distinct kinds of phenomena. For economics, it is human reason and agency; for the natural sciences, it is matter, force, and energy. Although, in principle, the nomotheticity of human agency might be reducible to physical brain states and, ultimately, quantum mechanics, there is little hope—on the horizon at least—that bridges will be found (see Dupré 1993). Economists are challenged enough to find the appropriate reductionist paths between their macroeconomic and their microeconomic theories. Furthermore, appeals to psychological regularities are few and far between in economics. The positivist legacy brought on by Paul Samuelson and others—to remain at the level of observation or "revealed preference"—is very pronounced and has hitherto set up substantial barriers to psychological probings, the advent of behavioral economics notwithstanding (see Hausman 1992, 22).

A concept of *the economy* that is severed from nature also lends itself to human control. A central motif of postwar economics is the ability to stabilize the economy via manipulations of the interest rate and the money supply. In short, many economists under the sway of Keynes believed that they could engineer the economy. A quick perusal of most macroeconomics textbooks of the 1970s and 1980s makes evident the strong commitment to "stabilization policy." Which "tools" to use are disputed, but not the overriding goal of stability. Exogenous and often unpredictable shocks like war or an oil crisis beset *the economy*, but economic planners are, through fiscal and monetary measures, able to steer us along a path of steady-state growth and stability. The rhetoric is very much like that of civil engineers. Wind shears, or earthquakes, or ordinary frost might threaten the stability of bridges, but engineers can overcome these shocks given the right plans. Macroeconomists convey a similar degree of confidence in achieving a national economy that couples low inflation and low unemployment with a healthy degree of economic growth. The laissez-faire stance of the Enlightenment has given way to one of engineering. Only with time will we be able to judge who had a deeper command of the path to human flourishing.

CHAPTER ONE

1. Two efforts to document the diffusion and ascent of economic ideas into government and the public sphere are Colander and Coats 1989 and Levy 1992.

2. Aristotle did not, as Polanyi once claimed, discover the economy, but he did observe many of its features (see Polanyi 1957; see also Finlay 1973; and Booth 1993).

3. When Smith refers to the *oeconomy* of England in *The Wealth of Nations*, he means the government's ability to be frugal and prudent (see, e.g., Smith 1776/1976, 1:457).

4. The *Encyclopedia of Philosophy* entry on "philosophical ideas of nature" makes much the same argument (see Hepburn 1967). The recent Routledge version of the same title has no such entry. Instead is found one on the nineteenth-century German movement known as *Naturphilosophie* (see Heidelberger 1998). On some recent efforts to define artifacts or the artificial, see Dennett 1990 and Hilpinen 1993.

5. Hankins illustrates this well with the juxtaposition of pictures, one a geometric garden by the landscaper for Versailles and the Tuileries, the other a Romantic impression of the eruption of Mount Vesuvius (see Hankins 1985, 4–5). A comparison of a Haydn symphony to Beethoven's Sixth captures a similar contrast between Enlightenment order and Romantic Sturm und Drang.

6. For Enlightenment images of nature, see Mukerji 1993. For Romantic images, see Knoepflmacher and Tennyson 1977.

7. Carl Becker called these "magic words," which "unobtrusively" come and go with the centuries (see Becker 1932, 47–48).

8. While it is true that Descartes inspired a mechanistic approach to the human body—La Mettrie's *l'homme machine*, e.g.—mechanism was fiercely contested by the

more dominant voice of vitalism. Geoffrey Sutton has argued that the influence of Cartesian dualism was far less significant during the eighteenth century than has generally been supposed, in part because Descartes himself believed the mind and body were not full-fledged dichotomies (see Sutton 1995, 180, 345).

9. This is true even of economic history (cliometrics) and the new institutionalism, insofar as both use the past as evidence to confirm neoclassical theory (see Schabas 1995b).

10. Some prominent works in this new genre are Mirowski 1989, Morgan 1990, and Weintraub 1991. I have addressed this transition in Schabas 1992, 2002.

11. For some conflicting definitions, see Sowell 1974, O'Brien 1975, and Hollander 1987. Brewer (1995) sees Turgot as the first classical economist.

12. Recently, Joel Kaye has argued that fourteenth-century economic thought profoundly reconfigured the conception of nature at the time, so there may well be a sense in which the causal chain runs in the other direction (see Kaye 1998). Arguably, seventeenth-century economic thought, particularly of the cameralist and mercantilist varieties, was at odds with the view that nature and economic processes were one and the same.

13. In my earlier work, I have also shown that Cairnes was more at one methodologically with Jevons than has traditionally been argued (see Schabas 1990b, 100–103).

14. On the historical roots of this, see, e.g., Kingsland 1994.

15. The subfield of environmental economics addresses the role of climate, but, in mainstream theory, climate rarely makes an appearance. There is some indication of a rekindling of interest in the role of the seasons (see, e.g., Kramer 1994; and Kamstra, Kramer, and Levi 2003). In statistical estimates, seasonal fluctuations are normally averaged out (see Morgan 1990).

16. Philippe Fontaine has argued that, for Turgot, there was a brand of "institutional individualism," but the kind of human agency that he identifies in Turgot does not undercut my claim here (Fontaine 1997).

17. Robert Heilbroner claims that "the whole objective of one part of [*The Theory of Moral Sentiments*] is to describe how the pressures of social judgment, finally internalized within the breast[,] . . . give rise to that epitome of socialized humanity, the prudent individual" (Heilbroner 1982, 429). So, while Heilbroner points to individuals in Smith's thought, they are always subsumed under broader social categories.

18. Alexander Rosenberg engages these issues in one of his earlier works (see Rosenberg 1976).

19. This claim is bolstered by findings in the philosophy of science. Since the work of Pierre Duhem and W. V. Quine, it has been widely accepted that empirical data underdetermine the theoretical content of a given science.

20. For extensive accounts of this period of economic theorizing, see Hutchison 1988 and Appleby 1978.

21. Needless to say, Locke's separation of the economic and political realms, at least developmentally, was itself a choice with strong political implications. In this sense,

Foucault was right to deem all discourses about wealth political, but the claim then becomes tautological.

22. The passage in question has to do with the shape of stars: "It has been shown that it is not in their nature to move themselves, and, since nature does nothing without reason or in vain, clearly she will have given things which possess no movement a shape particularly unadapted to movement. Such a shape is the sphere" (Aristotle 1984b, 480 [bk. 2, sec. 291b11–291b23]).

CHAPTER TWO

1. This claim is buttressed by my reading of Jevons's seminal works, where he initiates a program to treat economics as a branch of mechanics (see Schabas 1990b).

2. Kuhn traces these traditions back to Archimedes and Bacon, respectively, but sees Newton as the critical conjunction for the eighteenth century (see Kuhn 1977).

3. It was found, e.g., in Adam Smith's library, which contained a copy of Franklin's papers (see Mizuta 1967/2000, 95).

4. It was customary at the time for professors to write the doctoral theses of their students. The official author of this work is Isaac J. Biberg, but there is no challenge to the claim that the true author was Linnaeus. It was a widely circulated treatise.

5. For a detailed analysis of the *Nemesis divina,* see Lepenies 1982.

6. Linnaeus secretly dissented from this standard account of the Deluge. For more detail about the fanciful stories woven to make sense of this biblical episode, see Browne 1983.

7. Koerner also describes Linnaeus's attempts to induce abstention from imported luxuries. For example, Linnaeus circulated a popular almanac on the harmful effects of tea consumption, arguing that it made one stupid and feeble (Koerner 1999, 137). He also hoped to find substitutes for coffee, using burnt almonds, beans, and wheat. During his lifetime, Sweden often had a legal ban on coffee, presumably because it was a drain on its bullion (130–31).

8. For those who emphasize the similarity between Linnaeus and Darwin, see Stauffer (1960) and Worster (1977).

9. Schweber (1985) and Young (1985a) have made similar claims. Marx once remarked (in a June 18, 1862, letter to Engels) on the resemblance of Darwin's theory to classical political economy (see Ryazanskaya 1955/1975, 120).

10. As noted in chapter 1, there is some evidence to suggest that Lyell was well versed in political economy, particularly via his close friendships with George Poulett Scrope and Nassau Senior (see Rudwick 1974). Both Lyell and Darwin also attended a lecture by Richard Jones on the similarities between political economy and geology, at a meeting celebrating the thirtieth anniversary of the Geological Society (on the council of which Ricardo had once served). For more details on the many points of overlap between natural history and political economy in this period, see Rashid 1981b; Rudwick 1979; and Schweber 1977.

11. The movement known as *scientific creationism* is one of the most successful contemporary efforts to keep afloat a Christian scientific view (see Numbers 1986). Developments in twentieth-century physics, relativity theory, and quantum mechanics have also received a theological gloss, although mostly by nonphysicists (see Hiebert 1986).

12. I owe these categories, not to any one scholar, but to a collective effort to sort out the subject at a conference on the secularization of science that I coorganized with David Lindberg at the University of Wisconsin–Madison in 1989.

13. The widespread practice of shopping on Sunday, at least in North America, appears to be a clear sign that reverence for the temple has been outdone by the cash register. But this must also be squared with the fact that almost half of those shoppers, including a number of recent American presidents, still believe in some form of creationism as laid out in the Old Testament. As Chadwick (1975, 14) has observed, Lord Balfour played golf on Sundays, and Queen Victoria attended the theater during Lent, but that does not necessarily entail greater secularization.

14. In the 1740s, Abraham Tremblay discovered that a dismembered hydra could regenerate itself. This phenomenon was enthusiastically publicized by such materialists as Diderot, La Mettrie, and Buffon (see Hankins 1985, 131–33). On Pasteur's Christian agenda, see Geison 1995.

15. It is also possible to view Darwin's theory in a sacred light. Insofar as every living form was brought under a limited set of laws, albeit statistical ones, this could be seen to enhance God's wisdom in so arranging the universe. As Lovejoy (1936) showed, natural theology put much stock in the principles of plenitude and of continuity, to enhance God's omniscience. But, in many ways, the mechanisms of natural selection are just as finely wrought.

16. Paul Feyerabend (1975) has argued that science has become *the* religion of our modern age.

17. David Noble (1997) has argued that the major technological undertakings of the past century—NASA, the atomic bomb, and artificial intelligence—have been promoted mostly by Christian visionaries.

18. For arguments to the contrary, see Alston 1991.

19. "Nature and nature's laws lay hid in night: / God said, Let Newton be! And all was light" (Alexander Pope, "Epigraph Intended for Sir Isaac Newton").

20. On the advent of quantification in the world of science and industry, see T. Porter 1995 and Wise 1995. On the seepage of the scientific spirit into industrial production, see Mokyr 2002.

CHAPTER THREE

1. The classic work on France as a center of intellectual activity generally during the Enlightenment is Gay 1969, but, for essays specifically on science, see Rousseau and Porter 1980; Clark, Golinski, and Schaffer 1999; and Porter 2003.

2. Spary is quoting Rousseau's article "Economie" in vol. 5 of Diderot and d'Alembert's *Encyclopédie* (1751–72). On the last category, *l'économie rustique,* see Spary 2003, 21. For two partial assessments of Rousseau's political economy, see Larrère 1992, chap. 2; and Prieto 2004.

3. For overall assessments, see Weulersse 1910; Meek 1962; Fox-Genovese 1976; Larrère 1992; and "Mini-Symposium on Physiocracy" 2002.

4. Voltaire used his characteristic sarcasm in his *L'homme aux quarante écus* (1768), while Galiani used economic analysis in his *Dialogues sur le commerce des bléds* (1770) (see Perkins 1979, 326–27). Bonnot de Mably, who was Condillac's brother, attacked the theoretical core in his *Doutes proposés aux philosophes économistes sur l'ordre naturel et essentiel des sociétés politiques* (1768). For a general overview, see Hutchison 1988, 285.

5. For the argument that Quesnay had in mind a painting or picture, not a table in the sense of a scientific chart, see Charles 2003, 2004.

6. Quesnay subsequently included fishing and mining in his account of economic production.

7. Jacques Necker, an anti-Physiocratic minster of finance who emphasized the idea of a balance, also performed and wrote about his experiments with the Leyden jar (see Riskin 2002, 114n).

8. Of Turgot's proclivities, Gillispie wrote: "Not that he [Turgot] ever took himself for a scientist, although among his student memoirs are essays on cosmology and mechanics exhibiting a degree of comprehension high enough that he clearly could have done science had he wished" (1980, 7).

9. For two exceptions, see Jolink 1996 and Lallement 2000.

CHAPTER FOUR

1. Although Hume was traditionally portrayed as a skeptic, since the 1970s most scholars agree that a careful reading shows that he was not (see, e.g., Stroud 1977; and Baier 1991).

2. Roger Emerson has made a compelling case for strong links between Scottish moral philosophy and natural science during the Enlightenment (see Emerson 1990).

3. To his credit, Rotwein underscores the importance of treating Hume's essays in conjunction with the *Treatise,* contrary to the practice of other scholars (see Rotwein 1970, xviii–xxi). For additional evidence that Hume's "economic" insights are found beyond these nine essays, see, e.g., Young 1990; Skinner 1993b; Schabas 1994b; Berdell 1996; Gatch 1996; and Wennerlind 2001.

4. Haakonssen (1994) has argued that, for Hume, the most pressing problem was that of political stability and that his essays can be seen as addressing this concern from one angle or another. His economic analysis was, thus, motivated by political ends, a view that resonates with the thesis put forth by Hirschman (1977).

5. This interpretation of Hume is by no means an original one. It has been promoted, e.g., by Stroud (1977). For a more sophisticated unpacking of Hume on free will, see Russell 1995.

6. There is considerable debate as to whether Hume was an atheist. His criticisms of the church cost him an academic career, and he refused to take last rites while on his deathbed. Perhaps most telling are the challenges to biblical authority in the *Dialogues concerning Natural Religion*. But, for some scholars, he may have remained a theist for all that (see, e.g., Andre 1993).

7. Pitson (1993) notes in Hume's account some important differences between humans and animals, particularly with respect to our capacity for virtue, but argues that Hume's overall emphasis is on the similarities. Seidler concludes: "Hume's treatment of animals may thus be considered as part of a tradition of sceptical opposition to rationalistic claims about the superiority of man" (1977, 369). This tradition included Montaigne and Mandeville as well as Hume's contemporaries Lord Monboddo and Jean-Jacques Rousseau.

8. Hume does not, in fact, cite Montesquieu directly, but Chamley (1975) has argued that his was the theory that Hume set out to challenge. Hume maintained that there was great uniformity of character among the Chinese despite the wide range of geographic conditions under which they lived. The reason was their strong and persistent form of government, which has instilled great "similarity of manners" (Hume 1778/1985, 200–204).

9. For a detailed description of the extant library attached to Steuart's course at Edinburgh, see Barfoot 1990. Among the collection were works by Boyle, Newton, Hauksbee, 'sGravesande, and natural historians, such as John Ray, on the origins and evolution of the earth.

10. For the discussion of geometry, see Hume 1739–40/2000, 38–39. Roger Emerson has suggested to me that Hume's teacher was George Campbell. Had Hume been a year younger, he might have studied with Maclaurin, who arrived at the University of Edinburgh in 1725. On the two unpublished essays on mathematics, see Gossman 1960.

11. In an essay published in 1742, Hume noted that Newton's theory was still contested by philosophers on the Continent (see Hume 1778/1985, 121). This was a just observation, but hardly one in keeping with the adulations of some of his fellow Britons. Barfoot has also suggested that Hume was critical of Newton (see Barfoot 1990, 160–61). I am less persuaded by Capaldi (1975), who argued that Hume's program was deeply influenced by Newton.

12. For an interesting debate among Ian Hacking, Nancy Nersessian, and others on thought experiments, see Hull, Forbes, and Okruhlik 1993, 271–310.

13. John Law and George Berkeley made similar forays. Others at the time were inventive methodologists. William Petty was gifted in the task of measurement and approximation, Richard Cantillon in theoretical implication, and François Quesnay in

the cultivation of models. But these are best kept distinct from thought experiments so defined.

14. While exiled in France in 1750, Steuart expressed a longing to be at Goodtrees, his Scottish estate, where he would have had the company of David Hume and enjoyed the electrical and magnetic experiments of Robert Steuart (or Stewart), Hume's former professor of natural philosophy. See Steuart's "Biographical Sketch of Own Life" (Steuart 1794/1966, xxxvi). I wish to thank Professor Andrew Skinner for bringing this to my attention.

15. I do not wish to suggest that there is anything unwarranted about Hume's reasoning by analogy. Most scientific discoveries stem from the hybridization of different subject areas and the use of metaphoric and analogical reasoning. A good, up-to-date account of the rich array of such examples in economics and science more generally can be found in Lagueux 1999.

16. Hume discusses in some detail the process of banking and shows a clear understanding of the mechanism of fractional reserves, which, in a sense, are a kind of condensed money (see Hume 1778/1985, 319, 353).

17. A recent evaluation of the quantity theory of money, while once again paying homage to Hume, recognizes that we still lack a complete account of the chain of events as well as the exact temporal framework (see Blaug 1995).

18. As Laidler (1991, 299) observes: "There is, of course, no way of performing such an experiment on real world data. In economics, counter-factual experiments are necessarily conceptual, and it is precisely the impossibility of bringing empirical evidence directly and unambiguously to bear on questions concerning the feasibility of non-accommodative policy and, if feasible, upon its consequences for prices, which lies at the heart of the continuing controversy about the quantity theory."

19. In his correspondence with James Oswald, Hume gives some estimate of the time involved. He speculates that the influx of specie would first lead to the consumption of imported goods, this period of consumption lasting a year (presumably he had some figure to show that warehouses had inventories sufficient to last a year), and thereafter "increase the people and industry." Were there no real effects, the additional money stock would flow out over the course of a quarter century (see Rotwein 1970, 197–98).

20. I do not know when the term *monetary shock* was first coined, but there was a detailed effort by Morris Copeland to model the flow of money on electricity rather than water. Copeland posited batteries instead of reservoirs and wires instead of conduits and argued that the high velocity of electricity better captures the flow of money and the sense in which debt is simultaneously the creation of credit (see Copeland 1952). Karl Brunner and Allan Meltzer have argued that money is just information, and, since information is now stored and transmitted electronically, the analogy almost becomes redundant (see Brunner and Meltzer 1971).

21. Recent assessments of Hume's endorsement of the economic growth in Europe

during the early modern period can be found in Brewer 1995 and Berdell 1996. Berdell also points to Hume's appreciation for what we would now label *technological innovation.* Indeed, Franklin's lightning rod was one of the first cases of a scientific theory inspiring a practical device with clear economic benefits.

22. Paul Wood has made the same observation regarding Hume's writings on religion and also suggested that, in the *Enquiry,* there is evidence of "natural historical methods of description and classification in the science of the mind" (1989, 99).

23. For good overviews of Enlightenment natural history, see Hankins 1985, chap. 5, and Spary 2000.

24. Paul Wood, drawing on the work of Gladys Bryson and Andrew Skinner, has suggested that interest in natural history, and in Buffon most notably, rivaled interest in Newtonian physics among the intellectual elite of the Scottish Enlightenment (see Wood 1989). Simon Schaffer has explored the infusion of agrarian culture and natural history into Scottish moral philosophy (see Schaffer 1997).

25. Donald Livingston has argued that "Hume, like Vico, was working towards a reform in philosophy that takes history, not natural science, as the paradigm of knowledge" (1984, ix). But Livingston focused on Hume's historical writings and emphasized his gift for narrative. Hume's economic essays, on the other hand, already evince an analytic style that does not fit Livingston's thesis.

CHAPTER FIVE

1. Smith was most popular with the French, judging from the obituary notices (see Teichgraeber 1986; and Rothschild 2001).

2. The term *oeconomist* was common currency in the eighteenth century, but it meant a good manager of household resources or one who practices or advocates saving. The *Oxford English Dictionary* gives 1804 as the first time the term is used to mean a teacher or student of the subject of political economy (in the Earl of Lauderdale's *Inquiry into the Nature and Origin of Public Wealth*). The Physiocrats called themselves *les économistes,* and both Hume and Smith referred to them as such.

3. Smith had sixteen manuscript volumes burned right before his death. In keeping with his pared-down published record, there are only 179 letters by Smith extant. We know that over 50 are still missing, but, even then, that averages out to only four letters per year. Smith clearly played his cards close to his chest (see Tribe 1999).

4. The scholarship of Haakonssen (1981), Brown (1994), and Griswold (1999) has rekindled an interest in Smith's moral philosophy.

5. To become modern was costly, and Glasgow did not always have the financial resources. Ross infers that Smith had Oxford and Glasgow in mind when he drew a comparison between rich and poor universities in *The Wealth of Nations* (see Smith 1776/1976, 2:772–73; and Ross 1995, 55).

6. According to J. R. R. Christie: "Eighteenth-century Scotland's most committed ethereal scientist was the chemist and physiologist William Cullen" (Christie 1981, 86).

7. Kevin Brown has argued that Smith wrote the essay later than has been commonly supposed, probably in 1758 rather than prior to 1752. This would support the view that Smith knew Franklin's published scientific papers (see Brown 1992, 334). Brown's argument, however, is also consistent with the more likely scenario that Smith wrote and revised his essay over the entire decade.

8. I find Puro's distinctions overly pedantic. There are effectively three different usages: "that which is commonplace"; "that which is essential"; and "that which transpires without human agency." For each of these, Puro multiplies examples (see Puro 1992). Griswold looks at nature alongside of the natural and, thus, also brings in the distinction between the natural and the supernatural and the sense of nature as a purposeful whole (see Griswold 1999, 314–17). His categories are the more satisfactory.

9. Lorraine Daston observes that nature was personified as "Nature" when its authority was invoked and the descriptions recorded by naturalists were employed: "For Enlightenment sensibilities, authority required personification, and personification in turn required description. This description in turn drew heavily upon the practices of seeing and writing cultivated by naturalists" (Daston 2004, 126).

10. In a letter to Hume of 1773, Smith wrote: "As I have left the care of all my literary papers to you, I must tell you that except those which I carry along with me there are none worth the publishing, but a fragment of a great work which contains a history of the Astronomical Systems that were successively in fashion down to the time of Des Cartes. Whether that might not be published as a fragment of an intended juvenile work, I leave entirely to your judgement; tho [*sic*] I begin to suspect myself that there is more refinement than solidity in some parts of it" (Smith 1977, 168).

11. On the role of custom and resemblance, see Smith 1795/1980, 37–42. Hume arrives at the central claim that "all our reasonings concerning causes and effects are deriv'd from nothing but custom" (see Hume 1739–40/2000, 123). Resemblance for Hume is the first principle by which the mind associates one idea with another (see Hume 1739–40/2000, 13; see also Wightman 1980, 11).

12. These findings were the determination of the earth's curvature in the polar regions, the result of an expedition by Pierre-Louis Moreau de Maupertuis and Alexis Clairaut to Lapland in 1737; the solution of the lunar orbit by Clairaut, announced in 1749 and published in 1752, in terms of a series of approximations of the three-body problem (Smith 1795/1980, 100–101); and Edmund Halley's prediction, made much more precise by Clairaut, that a comet would return in 1758 or 1759. A footnote, added later, notes that Halley's prediction was confirmed and also that "the whole of this Essay was written previous to the date here mentioned" (Smith 1795/1980, 103).

13. Hetherington takes the fact that Smith referred to Descartes, not Newton, as a sign, not that he was uncomfortable with the final section, but that he had already drafted the material, a claim that fits with the evidence of the post-1758 footnote (see Hetherington 1983, 500). There is another plausible interpretation: Smith added the footnote soon after 1758 and then forgot to correct it when he added the section on Newton after 1773. This would imply that the Cartesian material was drafted fully, but

not the material on Newton. Still, it would be odd for Smith to have ended with Descartes and not with Newton, and we know that he had some understanding of the latter from his courses at Glasgow. I suspect that the full rendering of the Newton section required a more mature Smith, but the jury is still out as to whether this transpired before or after 1773.

14. Simon Schaffer explores the links between Smith's analysis of the division of labor and the widespread fascination with automata (see Schaffer 1999, 129–35).

15. As Thomas Kuhn points out, Copernicus eliminated the major epicycles that had accounted for retrograde motion under the geocentric system, but he was forced to retain minor epicycles as a device to bridge theory and observation, essentially because his orbits were circles, not ellipses. Only Kepler was able to dispense with the minor epicycle (see Kuhn 1957, 68).

16. I refer here to the often-quoted observation: "It is not from the benevolence of the butcher, the brewer, or the baker, that we expect our dinner, but from their regard to their own interest" (Smith 1776/1976, 1:26–27).

17. The verb *to fix* is of ancient vintage. The *Oxford English Dictionary* offers several definitions, one of which is in the immaterial sense of implanting sentiments and memories securely in the mind. It also notes that Desaguliers, the natural philosopher of the early eighteenth century, had spoken of fires passing over bodies without fixing to them.

18. Norton Wise and Crosbie Smith have shown that, in early-nineteenth-century analyses, there was a close reciprocity between the economist's concept of labor and the physicist's definition of work in protoenergetic terms (see Wise with Smith 1989–90). This conceptual trade may well hark back to Adam Smith.

19. Locke also uses the verb *to fix* in the context of property rights. For example: "The labour that was mine, removing them out of that common state they were in, hath fixed my property in them." Again: "As much as any one can make use of to any advantage of life before it spoils, so much he may by his labour fix a property in" (Locke 1764/1980, 20). But, in both sentences, it is the property that is being fixed, not the labor. Still, this may have been sufficient to prompt Smith to his own articulation of the concept of fixed labor.

20. Roderick Home traces the evolution of the efforts to understand the nervous fluid as a form of the electric fluid. The two were often conjoined in the late 1740s, an identification that helped inspire medical usages of electricity, e.g., the work of P. J. C. Mauduyt, N.-P. Le Dru, and the celebrated Anton Mesmer of mesmerism fame (see Sutton 1981). Albrecht von Haller was the pivotal figure in attempting to sever the two fluids, ca. 1760, although Galvani's famous dissection of a frog brought them back together. Nevertheless, the nervous fluid was always grouped with the collection of subtle fluids (see Home 1970).

21. Stefano Fiori links Smith to Maupertuis's and Buffon's appeals to invisible orders, most notably Buffon's *moule intérieure* (internal formal cause), which guides the individual development of organisms. This could prove to be an additional source for

Smith's distinction between nominal price and natural price and, as I argue here, his grander scheme of moral deception (see Fiori 2001, esp. 442).

22. Wightman makes this argument on the basis of the presence of Berkeley's and the absence of Hume's, ideas in the essay (see Wightman 1980, 15, 133–34).

23. "There is a grandeur in this view of life. . . . [W]hilst this planet has gone cycling on according to the fixed law of gravity, from so simple a beginning endless forms most beautiful and most wonderful have been, and are being, evolved" (Darwin 1859/1968, 659–60). Darwin was very much indebted to the eternalism of Hutton and the epochal thinking of Buffon, if not directly, then through Charles Lyell. It would be almost unimaginable for his closing remarks to have been made without all this spadework.

24. Given what little was known about the atomic substratum, Berkeley was justified in doubting the fanciful accounts of Descartes and Boyle. For a recent overview of eighteenth-century critiques of the mechanical philosophy, see Reill 2003.

25. Gordon Wood, drawing on Mandeville and others, argues: "The idea of deception became the means by which the Augustan Age closed the gaps that often seemed to exist between causes and effects. . . . This problem of deception was a source of continuing fascination in eighteenth-century Anglo-American culture" (Wood 1982, 425).

26. For a recent catalog of disputes, see Hill 2001.

27. Apparently, the French received his moral philosophy in this manner. Condorcet's wife, Sophy Grouchy, undertook a translation of *The Theory of Moral Sentiments* during the Revolution, intending it to secularize morality (see Rothschild 2001, 68; Forget 2001).

CHAPTER SIX

1. For those who emphasize discontinuity, see Hutchison (1978) and Berg (1980, 43). Hollander (1979), on the other hand, has emphasized the continuity between Smith and Ricardo.

2. For Herschel's praise for political economy, see Herschel 1830/1987, 72–73. On Whewell's extensive forays into economics, see Henderson 1996a. And, on the contributions of Babbage, see Romano 1982 and Ashworth 1994. Section F of the British Association for the Advancement of Science was initially only for the science of statistics, but most of the founding members were also active in political economy and contributed papers that were often on economic topics (see Henderson 1996b).

3. As William Buckland observed, the fortuitous proximity of iron ore, coal, and limestone in the British Midlands was not accidental. Rather, he argued, it was a sign of providential design, a gift intended to make Britain the richest country in the world (see Gillispie 1959, 200–201).

4. On Marshall's appreciation of Ricardo, see Schabas 1990b, 128, 171–72. Ricardo demonstrated that, at the margin, no rent is paid and, thus, that rent does not enter into

the cost of production and, hence, exchange value. But the more rigorous use of marginal analysis requires the comparison of two marginal variables, as in the equimarginal principle.

5. Milgate and Stimson (1991) have argued that Ricardo broke from Bentham and Mill's influence and cultivated his own distinct ideas on the subject of representative government.

6. Some of the more prominent commentators are Blaug (1958), Hollander (1979), and Caravale (1985). For an analysis of these debates, see Peach 1993.

7. Ideas similar to Malthus's had been expounded by William Petty, Bernard Mandeville, Benjamin Franklin, and Giammaria Ortes (see, e.g., Hartwick 1988).

8. Patricia James compiled a list of "authorities quoted or cited by Malthus in his *Essay*," and it is astoundingly long, running to at least two hundred entries (see Malthus 1803/1989, 2:253–57).

9. Malthus cites Hume—including his caveats regarding the "deceitful" status of the science of politics—in other passages (see, e.g., Malthus 1803/1989, 2:185).

10. Hollander argues that, by the 1820s, Malthus had moved away from this view, going so far as to endorse free agricultural trade, and came to appreciate Britain's manufacturing sector. But this was not the Malthus that people knew. Even those to whom he communicated his change of heart, namely, Nassau Senior, Thomas Chalmers, and Jane Marcet, did not publicize it (see Hollander 1997, 866).

11. Rashid has argued that Malthus's definition of *wealth* was driven by his goal of national income accounting. Material objects are measurable in a way that services are not, and I do not dispute that this was one consideration in shaping Malthus's definition. But I believe the more important consideration was a predilection for agrarian products that are clearly physical goods (see Rashid 1981a, 68–70).

12. Pullen suggests that Malthus's *Principles* made more of a mark than has generally been supposed (see Malthus 1820/1989, xviii). Rashid (1981a) and Winch (1996) make similar arguments.

13. Hollander has proposed that, prior to 1798, Malthus had a secondhand knowledge of Physiocracy, from Smith in particular. In the second edition of the *Essay* (1803), he cited both Dupont and Mirabeau and introduced several passages contrasting Smith with the Physiocrats. Hollander has argued that, by the 1803 edition, Malthus had become more doctrinaire than the Physiocrats but that, in his later years, after 1820, he began to relinquish their views (see Hollander 1997, 353, 404, 407, 866).

14. Malthus wrote of a poor, unemployed man: "[He] has no claim of *right* to the smallest portion of food, and, in fact, has no business to be where he is. At nature's mighty feast there is no vacant cover for him. She tells him to be gone, and will quickly execute her own orders, if he do not work on the compassion of some of her guests" (Malthus 1803/1989, 2:127). Pullen has suggested, although with hesitation, that Malthus toned down his view, in print at least, under pressure from theologians but that quite possibly his views remained unchanged from edition to edition. I find this to be improbable. As Hollander has shown, Malthus was quite inclined to reconsider his

position, and not one to kowtow to popular whim. I think that we can interpret the different positions in subsequent editions as just that, an ongoing evolution of his views on the subject. For Malthus's advocacy of moral restraint and the education of workers, see, e.g., Malthus 1803/1989, 2:241–45, 189–90, respectively.

15. Hollander notes that Paley favored population growth, so, in that sense, he and Malthus differed quite significantly (see Hollander 1997, 917).

16. Hollander noted one important difference; Mill did not advocate birth control after wedlock (see Hollander 1997, 948). Rashid, on very different grounds, has also suggested that Malthus's utilitarianism eclipsed his theological allegiances (see Rashid 1981a).

17. Hollander argues that Malthus made a firm break with protectionism and came to endorse liberal policies on trade and manufacturing (see Hollander 1997, 866).

18. In principle, Malthus too favored high wages for workers, but he also put more weight on religious education as the means to elevate humanity from the vicious cycle of dearth. He also believed that low wages would discourage marriage and childrearing and induce greater industry and, hence, prosperity.

19. Malthus, however, in opposition to Ricardo, maintained on the opening page of his *Principles* that political economy was more like morals than mathematics (see Malthus 1820/1989, 1:2).

20. For an opposing view, see White 1991.

21. I will draw my references from the most widely available edition, the fifth edition of 1864. But a study of the first and third editions shows that many of the key ideas that I here contrast to those of his contemporaries were already present in the 1825 version.

CHAPTER SEVEN

1. For appraisals of Mill's economics and its influence, see De Marchi 1974 and Hollander 1985.

2. On Mill in particular, see Whitaker 1975; Hausman 1981; Hollander 1983; and De Marchi 1986. On the more general theme of economics as physics, see Mirowski 1989; Ingrao and Israel 1990; and Schabas 1990b.

3. "If the laws of the production of all objects, or even of all material objects, which are useful or agreeable to mankind, were comprised in political economy, it would be difficult to say when the science would end: at the least, all or nearly all physical knowledge would be included in it" (Mill 1967, 314).

4. For an attempt to make sense of Mill's stand on consumption, see De Marchi 1972.

5. For a more detailed analysis of the physical and economic notions of work, see Wise with Smith 1989–90.

6. Sam Hollander informs me that Ricardo had already made this point, but not, it seems, as distinctively as Mill, who is able to carry it through in his analysis of rent.

7. Mill's stepdaughter, Helen Taylor, informs us in the "Introductory Notice" that Mill had intended to publish the essay in 1873 and that, uncharacteristically, he had made very few revisions to the original text, suggesting that his thoughts on the matter had "undergone no substantial change" (Mill 1969, 371–72).

8. Mill did not, however, subscribe to the "fanciful" pictures painted by Rousseau. In his view: "Savages are always liars. They have not the faintest notion of truth as a virtue" (Mill 1969, 395).

9. This is not to suggest that psychology became a science only in the nineteenth century. Recent scholarship maintains that "psychology as a natural science was not *invented* during the eighteenth century but *remade*" (Hatfield 1995, 188). Fox (1987) and Vidal (1993) have also stressed the coherence and appeal of psychology, or the science of the mind, during the Enlightenment. More recently, Goldstein (2003) has emphasized the role of Locke and Condillac.

10. Comin (2001) emphasizes the differences between Jevons and Wicksteed, the latter thinker being more drawn to realism and a commonsense approach to human behavior.

11. L. S. Jacyna has suggested that the Victorian science of the mind was part and parcel of a broader movement to unify nature and restore the moral foundation to scientific inquiry that had passed away with the demise of natural theology (Jacyna 1981, 129). This was probably less true for political economy, which seemed to ease into its secular state without significant challenges.

CHAPTER EIGHT

1. The fact that Mill's and Darwin's ideas were first drafted in the early 1840s suggests that their congeniality can be traced to a reliance on the same earlier sources— perhaps Linnaeus, Smith, Malthus, and Lyell?

2. Cambridge University, Marshall Library of Economics, Alfred Marshall Papers, item 6 (13), catalog no. 6/21/1.

3. On fitness, see Brandon and Beatty 1984.

4. Marshall praised Comte for his genius and for showing "how complex social phenomena are" (Marshall 1890/1920, 636), although he did not endorse Comte's view that economics had no right to a separate existence (see Pigou 1925, 163).

5. Although the essay is undated and Marshall never published it, he did read it before a meeting of the Grote Club in 1867. For a transcription, see Raffaelli 1994, 95–103.

6. According to John Maynard Keynes, Marshall expressed a belated wish to have devoted himself to psychology. It may have been this youthful insight into the complexity of the subject that initially steered him in other directions (see Pigou 1925, 37).

7. For a recent challenge to this view, particularly with respect to money, see Searle 1995.

8. As Hundert has observed, for Mandeville "the individual's point of view, while instinctive in naturally self-regarding creatures, in fact tends often to conceal the so-

cial significance of his actions" (Hundert 1994, 179). Hume repudiated the thesis that great men make significant innovations on the historical stage (see Hume 1778/1985, 476–77). Andy Denis has argued that Adam Smith "systematically denies the autonomy of the individual with respect to the whole of which he or she is part. For Smith, individual liberty is not the end, but the means, of sustaining social order and property" (Denis 1999, 71).

9. Hacking has helped us see the extent to which all physical constants, whether Avogadro's number or the gravitational constant, are subject to the apparatus of the laboratory and, thus, to human design and perception (see Hacking 1983, 1988a, 1991).

REFERENCES

Ahmad, Syed. 1990. "Adam Smith's Four Invisible Hands." *History of Political Economy* 22:137–44.

Alborn, Timothy. 1990. "Thomas Chalmers's Theology of Economics." In *Perspectives on the History of Economic Thought,* ed. Donald E. Moggridge, vol. 3. Aldershot: Edward Elgar.

Alston, William. 1991. *Perceiving God: The Epistemology of Religious Experience.* Ithaca, NY: Cornell University Press.

Andre, Shane. 1993. "Was Hume an Atheist?" *Hume Studies* 19:141–66.

Appleby, Joyce Oldham. 1978. *Economic Thought and Ideology in Seventeenth-Century England.* Princeton, NJ: Princeton University Press.

Aristotle. 1984a. *Nichomachean Ethics.* In *The Complete Works of Aristotle,* ed. Jonathan Barnes. Princeton, NJ: Princeton University Press.

———. 1984b. *On the Heavens.* In *The Complete Works of Aristotle,* ed. Jonathan Barnes. Princeton, NJ: Princeton University Press.

———. 1984c. *Politics.* In *The Complete Works of Aristotle,* ed. Jonathan Barnes. Princeton, NJ: Princeton University Press.

Armstrong, John. 1736. *The Oeconomy of Love.* London: T. Cooper.

Ashworth, William J. 1994. "The Calculating Eye: Baily, Herschel, Babbage and the Business of Astronomy." *British Journal for the History of Science* 27:409–41.

Atran, Scott. 1990. *Cognitive Foundations of Natural History: Toward an Anthropology of Science.* Cambridge: Cambridge University Press.

Ayer, Alfred Jules. 1936/1946. *Language, Truth and Logic.* 2nd ed. New York: Dover.

Bacon, Francis. 1605/1955. *The Advancement of Learning.* Edited by H. G. Dick. New York: Random House.

Baier, Annette. 1991. *A Progress of Sentiments: Reflections on Hume's Treatise.* Cambridge, MA: Harvard University Press.

Bain, Alexander. 1855/1868. *The Senses and the Intellect.* 3rd ed. London: Longmans, Green.

Baker, Keith Michael. 1964. "The Early History of the Term 'Social Science.'" *Annals of Science* 20:211–26.

———. 1975. *Condorcet: From Natural Philosophy to Social Mathematics.* Chicago: University of Chicago Press.

Banzhaf, Spencer H. 2000. "Productive Nature and the Net Product: Quesnay's Economics Animal and Political." *History of Political Economy* 32:517–51.

Banzhaf, Spencer H., and Neil De Marchi. 1994. "Images of Generation: Animal Oeconomy and Prosperity in Doctors Mandeville and Quesnay." Working paper, Duke University.

Barfoot, Michael. 1990. "Hume and the Culture of Science in the Early Eighteenth Century." In *Studies in the Philosophy of the Scottish Enlightenment,* ed. M. A. Stewart. Oxford: Clarendon.

Barnes, Barry, and Steven Shapin, eds. 1979. *Natural Order: Historical Studies of Scientific Culture.* Beverly Hills, CA: Sage.

Basalla, George. 1975. "Observations on the Present Status of the History of Science in the United States." *Isis* 66:467–70.

Bateman, Bradley. 1996. *Keynes's Uncertain Revolution.* Ann Arbor: University of Michigan Press.

Becker, Carl L. 1932. *The Heavenly City of the Eighteenth-Century Philosophers.* New Haven, CT: Yale University Press.

Becker, Gary S. 1976. *The Economic Approach to Human Behavior.* Chicago: University of Chicago Press.

———. 1996. *Accounting for Tastes.* Cambridge, MA: Harvard University Press.

Ben-David, Joseph. 1971. *The Scientist's Role in Society.* Chicago: University of Chicago Press.

Béraud, Alain, and Gilbert Faccarello, eds. 2000. *Nouvelle histoire de la pensée économique.* 3 vols. Paris: Editions la Découverte.

Berdell, John F. 1996. "Innovation and Trade: David Hume and the Case for Freer Trade." *History of Political Economy* 28:109–26.

Berg, Maxine. 1980. *The Machinery Question and the Making of Political Economy, 1815–1848.* Cambridge: Cambridge University Press.

Berkeley, George. 1734/1979. *Three Dialogues between Hylas and Philonous.* 3rd ed. Edited by Robert Merrihew Adams. Indianapolis: Hackett. The 1st edition appeared in 1713.

Black, R. D. Collison. 1990. "Jevons, Marshall and the Utilitarian Tradition." *Scottish Journal of Political Economy* 37:5–17.

Blaug, Mark. 1958. *Ricardian Economics.* New Haven, CT: Yale University Press.

———. 1972. "Was There a Marginal Revolution?" *History of Political Economy* 4:269–80.

———. 1978. *Economic Theory in Retrospect.* 3rd ed. Cambridge: Cambridge University Press.

———. 1995. "Why Is the Quantity Theory of Money the Oldest Surviving Theory in Economics?" In *The Quantity Theory of Money,* ed. Mark Blaug et al. Aldershot: Edward Elgar.

Blaug, Mark, et al. 1995. *The Quantity Theory of Money.* Aldershot: Edward Elgar.

Booth, William James. 1993. *Households: On the Moral Architecture of the Economy.* Ithaca, NY: Cornell University Press.

Bos, H. J. M. 1980. "Mathematics and Rational Mechanics." In *The Ferment of Knowledge: Studies in the Historiography of Eighteenth-Century Science,* ed. G. S. Rousseau and Roy Porter. Cambridge: Cambridge University Press.

Boulding, Kenneth E. 1981. *Evolutionary Economics.* Beverly Hills, CA: Sage.

Bowler, Peter J. 1989. *Evolution: The History of an Idea.* Berkeley and Los Angeles: University of California Press.

Brandon, Robert, and John Beatty. 1984. "The Propensity Interpretation of Fitness: No Interpretation Is No Substitute." *Philosophy of Science* 51:342–57.

Breton, Yves. 1998. "French Economists and Marginalism (1871–1918)." In *Studies in the History of French Political Economy: From Bodin to Walras,* ed. Gilbert Faccarello. New York: Routledge.

Brewer, Anthony. 1992. *Richard Cantillon: Pioneer of Economic Theory.* London: Routledge.

———. 1995. "The Concept of Growth in Eighteenth-Century Economics." *History of Political Economy* 27:609–38.

Brewer, John, and Roy Porter, eds. 1993. *Consumption and the World of Goods.* London: Routledge.

Brooke, John Hedley. 1991. *Science and Religion: Some Historical Perspectives.* Cambridge: Cambridge University Press.

———. 2003. "Science and Religion." In *Eighteenth-Century Science (The Cambridge History of Science,* vol. 4), ed. Roy Porter. Cambridge: Cambridge University Press.

Brown, Kevin L. 1992. "Dating Adam Smith's Essay 'Of the External Senses.'" *Journal of the History of Ideas* 53:333–37.

Brown, Robert. 1984. *The Nature of Social Laws, Machiavelli to Mill.* Cambridge: Cambridge University Press.

Brown, Theodore M. 1968. "The Mechanical Philosophy and the 'Animal Oeconomy'—a Study in the Development of English Physiology in the Seventeenth and Early Eighteenth Century." Ph.D. diss., Princeton University.

Brown, Vivienne. 1994. *Adam Smith's Discourse: Canonicity, Commerce and Conscience.* London: Routlege.

Browne, Janet. 1983. *The Secular Ark: Studies in the History of Biogeography.* New Haven, CT: Yale University Press.

Brunner, Karl, and Allan H. Meltzer. 1971. "The Uses of Money: Money in the Theory of an Exchange Economy." *American Economic Review* 61:784–805.

Bryson, Gladys. 1945. *Man and Society: The Scottish Inquiry of the Eighteenth Century.* Princeton, NJ: Princeton University Press.

Buchwald, Jed Z. 1985. *From Maxwell to Microphysics.* Chicago: University of Chicago Press.

Buffon, George Leclerc de. 1749–76. *Histoire naturelle, générale et particulière.* 15 vols. Paris: De l'imprimerie royale.

Burrow, J. W. 1966. *Evolution and Society: A Study in Victorian Social Theory.* Cambridge: Cambridge University Press.

Cahan, David, ed. 2003. *From Natural Philosophy to the Sciences: Writing the History of Nineteenth-Century Science.* Chicago: University of Chicago Press.

Cairnes, John Elliott. 1875/1965. *The Character and Logical Method of Political Economy.* Reprint, New York: Augustus M. Kelley.

Campbell, T. D. 1975. "Scientific Explanation and Ethical Justification in the *Moral Sentiments.*" In *Essays on Adam Smith,* ed. Andrew Skinner and Thomas Wilson. Oxford: Clarendon.

Cantillon, Richard. 1755/1964. *Essai sur la nature du commerce en général.* Translated by H. Higgs. Reprint, New York: Augustus M. Kelley.

Capaldi, Nicholas. 1975. *David Hume: The Newtonian Philosopher.* Boston: Twayne.

———. 1976. "Hume's Theory of the Passions." In *Hume: A Re-Evaluation,* ed. Donald W. Livingston and James T. King. New York: Fordham University Press.

Caravale, Giovanni, ed. 1985. *The Legacy of Ricardo.* Oxford: Basil Blackwell.

Cardoso, José Luís. 2003. "From Natural History to Political Economy: The Enlightened Mission of Domenico Vandelli in Late Eighteenth-Century Portugal." *Studies in History and Philosophy of Science* 34:781–803.

Carrithers, David. 1995. "The Enlightenment Science of Society." In *Inventing Human Science: Eighteenth-Century Domains,* ed. Christopher Fox, Roy Porter, and Robert Wokler. Berkeley and Los Angeles: University of California Press.

Cartwright, Nancy. 1983. *How the Laws of Physics Lie.* Oxford: Oxford University Press.

———. 1989. *Nature's Capacities and Their Measurement.* Oxford: Oxford University Press.

———. 1999. *The Dappled World: A Study of the Boundaries of Science.* Cambridge: Cambridge University Press.

Chadwick, Owen. 1975. *The Secularization of the European Mind in the Nineteenth Century.* Cambridge: Cambridge University Press.

Chaigneau, Nicolas. 1995. "A Sketch in the History of Indifference Curves: Edgeworth, Fisher and the Role of Psychology." Unpublished manuscript, Groupe de Recherches en Epistémologie de la Science Economique, Université de Paris I.

Chamley, Paul E. 1975. "The Conflict between Montesquieu and Hume." In *Essays on Adam Smith,* ed. Andrew Skinner and T. Wilson. Oxford: Clarendon.

Charles, Loïc. 2003. "The Visual History of the *Tableau Économique.*" *European Journal of the History of Economic Thought* 10:527–50.

———. 2004. "The *Tableau Économique* as Rational Recreation." *History of Political Economy* 36:445–74.

———. In press. "French Cultural Politics and the Dissemination of Hume's *Political Discourses* on the Continent (1750–1770)." In *Essays on David Hume's Political Economy,* ed. Carl Wennerlind and Margaret Schabas. London: Routledge.

Chasse, John Dennis. 1984. "Marshall, the Human Agent and Economic Growth: Wants and Activities Revisited." *History of Political Economy* 16:381–404.

Checkland, S. G. 1951. "The Advent of Academic Economics in England." *Manchester School* 19:43–70.

Christensen, Paul P. 1994. "Fire, Motion, and Productivity: The Proto-Energetics of Nature and Economy in François Quesnay." In *Natural Images in Economic Thought,* ed. Philip Mirowski. Cambridge: Cambridge University Press.

———. 2003. "Epicurean and Stoic Sources for Boisguilbert's Physiological and Hippocratic Vison of Nature and Economics." In *Oeconomies in the Age of Newton,* ed. Margaret Schabas and Neil De Marchi. Durham, NC: Duke University Press.

Christie, J. R. R. 1981. "Ether and the Science of Chemistry: 1740–1790." In *Conceptions of Ether,* ed. G. N. Cantor and M. J. S. Hodge. Cambridge: Cambridge University Press.

Clark, Charles. 1992. *Economic Theory and Natural Philosophy.* Aldershot: Edward Elgar.

Clark, William, Jan Golinski, and Simon Schaffer, eds. 1999. *The Sciences in Enlightened Europe.* Chicago: University of Chicago Press.

Cleaver, K. C. 1989. "Adam Smith on Astronomy." *History of Science* 27:211–18.

Clive, John. 1970. "The Social Background of the Scottish Renaissance." In *Scotland in the Age of Improvement,* ed. N. T. Phillipson and Rosalind Mitchison. Edinburgh: Edinburgh University Press.

Coats, A. W. 1968. "The Origins and Early Development of the Royal Economic Society." *Economic Journal* 78:349–71.

———. 1972. "The Economic and Social Context of the Marginal Revolution of the 1870's." *History of Political Economy* 4:303–24.

———. 1990. "Marshall and Ethics." In *Alfred Marshall in Retrospect,* ed. Rita McWilliams Tullberg. Aldershot: Edward Elgar.

Cohen, I. Bernard. 1990. *Benjamin Franklin's Science.* Cambridge, MA: Harvard University Press.

———. 1994. "Preface and Introduction." In *The Natural Sciences and the Social Sciences: Some Critical and Historical Perspectives,* ed. I. Bernard Cohen. Dordrecht: Kluwer.

Colander, David, and A. W. Coats, eds. 1989. *The Spread of Economic Ideas.* Cambridge: Cambridge University Press.

Coleman, William. 1982. *Death Is a Social Disease.* Madison: University of Wisconsin Press.

Collingwood, R. G. 1945. *The Idea of Nature.* New York: Oxford University Press.

Collini, Stefan, Donald Winch, and John Burrow. 1983. *That Noble Science of Politics.* Cambridge: Cambridge University Press.

Comin, Flavio. 2001. "Jevons and Wicksteed: Crossing Borders in the History of Economics." In *Economics and Interdisciplinary Exchange,* ed. Guido Erreygers. London: Routledge.

Condillac, Étienne Bonnot de. 1776/1997. *Commerce and Government.* Translated by Shelagh Eltis. Cheltenham: Edward Elgar.

Cook, Harold J. 1999. "Bernard Mandeville and the Therapy of 'the Clever Politician.'" *Journal of the History of Ideas* 60:101–24.

Copeland, Morris. 1952. *A Study of Moneyflows in the United States.* New York: National Bureau of Economic Research.

Cournot, A. A. 1838/1960. *Researches into the Mathematical Principles of the Theory of Wealth.* Translated by N. T. Bacon. Reprint, New York: Augustus M. Kelley.

Creedy, John. 1986. *Edgeworth and the Development of Neoclassical Economics.* Oxford: Basil Blackwell.

Cropsey, Joseph. 1975. "Adam Smith and Political Philosophy." In *Essays on Adam Smith,* ed. Andrew Skinner and T. Wilson. Oxford: Clarendon.

Darnton, Robert. 1985. *The Great Cat Massacre and Other Episodes in French Cultural History.* New York: Vintage.

Darwin, Charles. 1859/1968. *The Origin of Species by Means of Natural Selection.* London: John Murray. Reprint, Harmondsworth: Penguin.

Daston, Lorraine. 1978. "British Responses to Psycho-Physiology, 1860–1900." *Isis* 69:192–208.

———. 1982. "The Theory of Will versus Science of Mind." In *The Problematic Science: Psychology in Nineteenth-Century Thought,* ed. William R. Woodward and Mitchell G. Ash. New York: Praeger.

———. 1988. *Classical Probability in the Enlightenment.* Princeton, NJ: Princeton University Press.

———. 1998. "The Nature of Nature in Early Modern Europe." *Configurations* 6:149–72.

———. 2004. "Attention and the Values of Nature in the Enlightenment." In *The Moral Authority of Nature,* ed. Lorraine Daston and Fernando Vidal. Chicago: University of Chicago Press.

Daston, Lorraine, and Fernando Vidal, eds. 2004. *The Moral Authority of Nature.* Chicago: University of Chicago Press.

Davie, George. 1979. "Berkeley, Hume, and the Central Problem of Scottish Philosophy." In *McGill Hume Studies,* ed. D. F. Norton, N. Capaldi, and W. L. Robison. San Diego: Austin Hill.

Davis, J. Ronnie. 1990. "Adam Smith on the Providential Reconciliation of Individual and Social Interests: Is Man Led by an Invisible Hand or Misled by a Sleight of Hand?" *History of Political Economy* 22:341–52.

Davis, John B. 1994. *Keynes's Philosophical Development.* Cambridge: Cambridge University Press.

———. 2003. *The Theory of the Individual in Economics: Identity and Value.* London: Routledge.

Defoe, Daniel. 1730. *A Brief State of the Inland or Home Trade of England and of the Oppressions It Suffers and the Dangers Which threaten It from the Invasion of Hawkers, Pedlars, and Clandestine Traders of All Sorts.* London: Longmans, Green.

De Marchi, Neil B. 1972. "Mill and Cairnes and the Emergence of Marginalism in England." *History of Political Economy* 4:344–63.

———. 1973. "The Noxious Influence of Authority: A Correction of Jevons's Charge." *Journal of Law and Economics* 16:179–89.

———. 1974. "The Success of Mill's Principles." *History of Political Economy* 6:119–57.

———. 1986. "Mill's Unrevised Philosophy of Economics: A Comment on Hausman." *Philosophy of Science* 53:89–100.

Denis, Andy. 1999. "Was Adam Smith an Individualist?" *History of the Human Sciences* 12:71–86.

Dennett, Daniel. 1990. "The Interpretation of Texts, People and Other Artifacts." *Philosophy and Phenomenological Research* 1 (suppl.): 177–94.

d'Holbach, Paul Henri. 1781/1990. *Système de la nature.* 2 vols. 2nd ed. Reprint, Paris: Fayard. The 1st edition appeared in 1770.

Dickey, Laurence. 1986. "Historicizing the 'Adam Smith Problem': Conceptual, Historiographical, and Textual Issues." *Journal of Modern History* 58:579–609.

Diderot, Dénis. 1753/1981. *De l'interpretation de la nature.* Vol. 9 of *Diderot oeuvres complètes.* Paris: Hermann.

Dobb, Maurice. 1973. *Theory of Value and Distribution since Adam Smith: Ideology and Economic Theory.* Cambridge: Cambridge University Press.

Donovan, Arthur. 1975. *Philosophical Chemistry in the Scottish Enlightenment: The Doctrines and Discoveries of William Cullen and Joseph Black.* Edinburgh: University of Edinburgh Press.

———. 1976. "Pneumatic Chemistry and Newtonian Natural Philosophy in the Eighteenth Century: William Cullen and Joseph Black." *Isis* 67:217–28.

Drayton, Richard. 2000. *Nature's Government: Science, Imperial Britain, and the "Improvement" of the World.* New Haven, CT: Yale University Press.

Duke, Michael. 1979. "David Hume and Monetary Adjustment." *History of Political Economy* 11:572–87.

Dumez, Hervé. 1985. *L'economiste, la science et le pouvoir: Le cas Walras.* Paris: Presses Universitaires de France.

Dunn, John. 1980. *The Political Thought of John Locke.* Cambridge: Cambridge University Press.

———. 1983. "From Applied Theology to Social Analysis: The Break between John Locke and the Scottish Enlightenment." In *Wealth and Virtue,* ed. Istvan Hont and Michael Ignatieff. Cambridge: Cambridge University Press.

Dupré, John. 1993. *The Disorder of Things: Metaphysical Foundations of the Disunity of Science.* Cambridge, MA: Harvard University Press.

Durant, John R. 1985. "The Ascent of Nature in Darwin's *Descent of Man.*" In *The Darwinian Heritage,* ed. David Kohn. Princeton, NJ: Princeton University Press.

Edgeworth, Francis Ysidro. 1881/1967. *Mathematical Psychics: An Essay on the Application of Mathematics to the Moral Sciences.* Reprint, New York: Augustus M. Kelley.

———. 1885/1994. "The Calculus of Probabilities Applied to Psychical Research I." In *Edgeworth on Chance, Economic Hazard, and Statistics,* ed. Philip Mirowski. Lanham, MD: Rowman & Littlefield.

Ege, Ragip. 2000. "L'économie politique et les systèmes sociaux pendant la premiere moitié du xixe siècle." In *Nouvelle histoire de la pensée économique,* ed. Alain Béraud and Gilbert Faccarello, vol. 2. Paris: Editions la Découverte.

Egerton, Frank. 1973. "Changing Concepts of the Balance of Nature." *Quarterly Review of Biology* 48:322–49.

Eisley, Loren. 1961. *Darwin's Century.* New York: Doubleday.

Ekelund, Robert B., Jr. and Robert F. Hébert. 1999. *The Secret Origins of Modern Microeconomics.* Chicago: University of Chicago Press.

Eltis, Walter. 1995. "L'Abbé de Condillac and the Physiocrats." *History of Political Economy* 27:217–36.

Emerson, Roger. 1990. "Science and Moral Philosophy in the Scottish Enlightenment." In *Studies in the Philosophy of the Scottish Enlightenment,* ed. M. A. Stewart. Oxford: Clarendon.

Evensky, Jerry. 1987. "The Two Voices of Adam Smith: Moral Philosopher and Social Critic." *History of Political Economy* 19:447–68.

Faccarello, Gilbert, ed. 1998. *Studies in the History of French Political Economy: From Bodin to Walras.* London: Routledge.

———. 1999. *The Foundation of Laissez-Faire: The Economics of Pierre de Boisguilbert.* London: Routledge.

Feyerabend, Paul. 1975. *Against Method.* London: Verso.

Finlay, Moses I. 1973. *The Ancient Economy.* Berkeley and Los Angeles: University of California Press.

Fiori, Stefano. 2001. "Visible and Invisible Order: The Theoretical Duality of

Smith's Political Economy." *European Journal of the History of Economic Thought* 8:429–48.

Foley, Duncan. 1987. "Money in Economic Activity." In *Money: The New Palgrave,* ed. John Eatwell, Murray Milgate, and Peter Newman. New York: Norton.

Foley, Vernard. 1973. "An Origin of the *Tableau Économique.*" *History of Political Economy* 5:121–50.

———. 1976. *The Social Physics of Adam Smith.* West Lafayette, IN: Purdue University Press.

Fontaine, Philippe. 1992. "Social Progress and Economic Behaviour in Turgot." In *Perspectives on the History of Economic Thought,* vol. 7, ed. S. Todd Lowry. Aldershot: Edward Elgar.

———. 1997. "Turgot's 'Institutional Individualism.'" *History of Political Economy* 29:1–20.

Fontenelle, Bernard le Bovier de. 1686/1990. *Conversations on the Plurality of Worlds.* Translated by H. A. Hargreaves. Berkeley and Los Angeles: University of California Press.

Forbes, Duncan. 1975. *Hume's Philosophical Politics.* Cambridge: Cambridge University Press.

Force, James E. 1987. "Hume's Interest in Newton and Science." *Hume Studies* 13:166–216.

Forget, Evelyn L. 1999. *The Social Economics of Jean-Baptiste Say: Markets and Virtue.* London: Routledge.

———. 2001. "Cultivating Sympathy: Sophie Condorcet's Letters on Sympathy." *Journal of the History of Economic Thought* 23:319–37.

———. 2003. "Evocations of Sympathy: Sympathetic Imagery in Eighteenth-Century Social Theory and Physiology." In *Oeconomies in the Age of Newton,* ed. Margaret Schabas and Neil De Marchi. Durham, NC: Duke University Press.

Foucault, Michel. 1970. *The Order of Things: An Archeology of the Human Sciences.* New York: Vintage.

Fox, Christopher. 1987. "Defining Eighteenth-Century Psychology: Some Problems and Perspectives." In *Psychology and Literature in the Eighteenth Century,* ed. Christopher Fox. New York: AMS.

Fox, Christopher, Roy Porter, and Robert Wokler, eds. 1995. *Inventing Human Science: Eighteenth-Century Domains.* Berkeley and Los Angeles: University of California Press.

Fox-Genovese, Elizabeth. 1976. *The Origins of Physiocracy.* Ithaca, NY: Cornell University Press.

Foxwell, Herbert Somerton. 1887. "The Economic Movement in England." *Quarterly Journal of Economics* 2:84–103.

Franklin, Benjamin. 1754/1968. *New Experiments and Observations on Electricity.*

2nd ed. In *Steps in the Scientific Tradition,* ed. Richard S. Westfall and Victor E. Thoren. New York: John Wiley. Parts of *New Experiments* first circulated in 1747 in the form of two letters to Peter Collinson and subsequently appeared in print, with supplementary material of 1749, in the *Philosophical Transactions of the Royal Society* of 1751.

———. 1762/1771. "Letter from Dr. B. Franklin to D. Hume, Esq., on the Method of Securing Houses from the Effects of Lightning." In *Essays and Observations, Physical and Literary, of the Philosophical Society of Edinburgh,* vol. 3. Edinburgh: John Balfour.

Frye, Northrop. 1990–91. "Varieties of Eighteenth-Century Sensibility." *Eighteenth-Century Studies* 24:157–72.

Funkenstein, Amos. 1986. *Theology and the Scientific Imagination from the Middle Ages to the Seventeenth Century.* Princeton, NJ: Princeton University Press.

Garber, Daniel. 1995. "Experiment, Community, and the Constitution of Nature in the Seventeenth Century." *Perspectives on Science* 3:173–203.

Gascoigne, John. 2003. "Ideas of Nature: Natural Philosophy." In *Eighteenth-Century Science* (*The Cambridge History of Science,* vol. 4), ed. Roy Porter. Cambridge: Cambridge University Press.

Gatch, Loren. 1996. "To Redeem Metal with Paper: David Hume's Philosophy of Money." *Hume Studies* 22:169–91.

Gauthier, David. 1992. "Artificial Virtues and the Sensible Knave." *Hume Studies* 18:401–28.

Gay, Peter. 1969. *The Enlightenment: An Interpretation.* New York: Norton.

Geison, Gerald. 1995. *The Private Science of Louis Pasteur.* Princeton, NJ: Princeton University Press.

Gillispie, Charles Coulston. 1959. *Genesis and Geology.* New York: Harper & Row.

———. 1960. *The Edge of Objectivity.* Princeton, NJ: Princeton University Press.

———. 1980. *Science and Polity in France at the End of the Old Regime.* Princeton, NJ: Princeton University Press.

Goldstein, Jan. 2003. "Bringing the Psyche into Scientific Focus." In *The Modern Social Sciences* (*The Cambridge History of Science,* vol. 7), ed. Theodore M. Porter and Dorothy Ross. Cambridge: Cambridge University Press.

Golinski, Jan. 1992. *Science as Public Culture.* Cambridge: Cambridge University Press.

———. 1999. "Barometers of Change." In *The Sciences in Enlightened Europe,* ed. William Clark, Jan Golinski, and Simon Schaffer. Chicago: University of Chicago Press.

Goodhart, C. A. E. 1984. *Monetary Theory and Practice.* London: Macmillan.

Gordon, Barry. 1976. *Political Economy in Parliament, 1819–1823.* London: Macmillan.

Gordon, H. Scott. 1973. "Alfred Marshall and the Development of Economics as a

Science." In *Foundations of Scientific Method: The Nineteenth Century,* ed. R. N. Giere and R. S. Westfall. Bloomington: Indiana University Press.

———. 1989. "Darwin and Political Economy: The Connection Reconsidered." *Journal of the History of Biology* 22:437–59.

Gossman, Lionel. 1960. "Two Unpublished Essays on Mathematics in the Hume Papers." *Journal of the History of Ideas* 21:442–49.

Grant, Edward. 1986. "Science and Theology in the Middle Ages." In *God and Nature: Historical Essays on the Encounter between Christianity and Science,* ed. David C. Lindberg and Ronald L. Numbers. Berkeley and Los Angeles: University of California Press.

Griswold, Charles. 1999. *Adam Smith and the Virtues of Enlightenment.* Cambridge: Cambridge University Press.

Groenewegen, P. D. 1977. *The Economics of A. R. J. Turgot.* The Hague: Martinus Nijhoff.

———. 1983. "Turgot's Place in the History of Economic Thought: A Bicentenary Estimate." *History of Political Economy* 15:585–616.

———. 1990. "Marshall and Hegel." *Economie Appliquée* 43:63–84.

Haakonssen, Knud. 1981. *The Science of a Legislator: The Natural Jurisprudence of David Hume and Adam Smith.* Cambridge: Cambridge University Press.

———. 1994. Introduction to *David Hume: Political Essays,* ed. Knud Haakonssen. Cambridge: Cambridge University Press.

———. 1998. Introduction to *Adam Smith,* ed. Knud Haakonssen. Brookfield: Ashgate.

Hacking, Ian. 1983. *Representing and Intervening.* New York: Cambridge University Press.

———. 1988a. "On the Stability of the Laboratory Sciences." *Journal of Philosophy* 85:507–14.

———. 1988b. "Telepathy: Origins of Randomization and the Design of Experiments." *Isis* 79:427–51.

———. 1991. "Artificial Phenomena." *British Journal for the History of Science* 23:235–41.

Hales, Stephen. 1727/1969. *Vegetable Staticks.* Reprint, New York: American Elsevier.

Halévy, Elie. 1972. *The Growth of Philosophic Radicalism.* Translated by Mary Morris. New York: Augustus M. Kelley.

Hankins, Thomas L. 1970. *Jean d'Alembert: Science and the Enlightenment.* Oxford: Oxford University Press.

———. 1985. *Science and the Enlightenment.* Cambridge: Cambridge University Press.

Hardin, Garret. 1960. "The Competitive Exclusion Principle." *Science* 131:1292–97.

Harman, P. M. 1982. *Energy, Force, and Matter: The Conceptual Development of Nineteenth-Century Physics.* Cambridge: Cambridge University Press.

Hartwick, John M. 1988. "Robert Wallace and Malthus and the Ratios." *History of Political Economy* 20:357–79.

Harvey-Phillips, M. B. 1984. "Malthus' Theodicy: The Intellectual Background of His Contribution to Political Economy." *History of Political Economy* 16:591–608.

Haskell, Thomas. 1977. *The Emergence of Professional Social Science.* Urbana: University of Illinois Press.

Hatfield, Gary. 1995. "Remaking the Science of Mind: Psychology as Natural Science." In *Inventing Human Science: Eighteenth-Century Domains,* ed. Christopher Fox, Roy Porter, and Robert Wokler. Berkeley and Los Angeles: University of California Press.

Hausman, Daniel. 1981. "John Stuart Mill's Philosophy of Economics." *Philosophy of Science* 48:383–85.

———. 1992. *The Inexact and Separate Science of Economics.* Cambridge: Cambridge University Press.

Hearn, Jeff. 1991. "Gender: Biology, Nature, and Capitalism." In *The Cambridge Companion to Marx,* ed. Terrell Carver. Cambridge: Cambridge University Press.

Heidelberger, M. 1998. "Naturphilosophie." In *Routledge Encyclopedia of Philosophy,* ed. E. Craig. London: Routledge.

Heilbron, John. 1979. *Electricity in the Seventeenth and Eighteenth Centuries: A Study of Early Modern Physics.* Berkeley and Los Angeles: University of California Press.

———. 1980. "Experimental Natural Philosophy." In *The Ferment of Knowledge: Studies in the Historiography of Eighteenth-Century Science,* ed. G. S. Rousseau and Roy Porter. Cambridge: Cambridge University Press.

Heilbroner, Robert L. 1982. "The Socialization of the Individual in Adam Smith." *History of Political Economy* 14:427–39.

Heimann, P. M. 1972. "The *Unseen Universe:* Physics and the Philosophy of Nature in Victorian Britain." *British Journal for the History of Science* 6:73–79.

———. 1978. "Voluntarism and Immanence: Conceptions of Nature in Eighteenth-Century Thought." *Journal of the History of Ideas* 39:271–83.

Henderson, James P. 1996a. *Early Mathematical Economics: William Whewell and the British Case.* Lanham, MD: Rowman & Littlefield.

———. 1996b. "Emerging Learned Societies: Economic Ideas in Context." *Journal of the History of Economic Thought* 18:186–206.

Hepburn, Ronald W. 1967. "Philosophical Ideas of Nature." In *The Encyclopedia of Philosophy,* ed. Paul Edwards. New York: Macmillan.

Herschel, John. 1830/1987. *Preliminary Discourse on the Study of Natural Philosophy.* Reprint, Chicago: University of Chicago Press.

Hetherington, Norriss S. 1983. "Isaac Newton's Influence on Adam Smith's Natural Laws in Economics." *Journal of the History of Ideas* 44:497–505.

Hiebert, Erwin N. 1986. "Modern Physics and Christian Faith." In *God and Nature: Historical Essays on the Encounter between Christianity and Science,* ed. David C. Lindberg and Ronald L. Numbers. Berkeley and Los Angeles: University of California Press.

Hill, Lisa. 2001. "The Hidden Theology of Adam Smith." *European Journal of the History of Economic Thought* 8:1–29.

Hilpinen, Risto. 1993. "Authors and Artifacts." *Proceedings of the Aristotelian Society* 43:155–78.

Hilton, Boyd. 1988. *The Age of Atonement: The Influence of Evangelicalism on Social and Economic Thought, 1795–1865.* Oxford: Clarendon.

Hirschleifer, Jack. 1977. "Economics from a Biological Viewpoint." *Journal of Law and Economics* 20:1–52.

Hirschman, Albert O. 1977. *The Passions and the Interests.* Princeton, NJ: Princeton University Press.

Hobsbawm, E. J. 1975. *The Age of Capital, 1848–1875.* London: Abacus.

Hodgson, Geoffrey M. 1993. *Economics and Evolution: Bringing Life Back into Economics.* Ann Arbor: University of Michigan Press.

Hollander, Samuel. 1973. *The Economics of Adam Smith.* Toronto: University of Toronto Press.

———. 1979. *The Economics of David Ricardo.* Toronto: University of Toronto Press.

———. 1983. "William Whewell and John Stuart Mill on the Methodology of Political Economy." *Studies in the History and Philosophy of Science* 4:127–68.

———. 1985. *The Economics of John Stuart Mill.* 2 vols. Toronto: University of Toronto Press.

———. 1986. "On Malthus's Population Principle and Social Reform." *History of Political Economy* 18:187–235.

———. 1987. *Classical Economics.* Oxford: Basil Blackwell.

———. 1997. *The Economics of Thomas Robert Malthus.* Toronto: University of Toronto Press.

Home, Roderick W. 1970. "Electricity and the Nervous Fluid." *Journal of the History of Biology* 3:235–51.

Hont, Istvan. 1983. "The 'Rich Country-Poor Country' Debate in Scottish Classical Political Economy." In *Wealth and Virtue,* ed. Istvan Hont and Michael Ignatieff. Cambridge: Cambridge University Press.

Hont, Istvan, and Michael Ignatieff, eds. 1983. *Wealth and Virtue.* Cambridge: Cambridge University Press.

Hull, David, Micky Forbes, and Kathleen Okruhlik, eds. 1993. *PSA 1992: Proceedings of the 1992 Biennial Meeting of the Philosophy of Science Association.* Vol. 2, *Symposia and Invited Papers.* East Lansing, MI: Philosophy of Science Association.

Hume, David. 1739–40/2000. *A Treatise of Human Nature.* Edited by David Fate Norton and Mary J. Norton. Oxford: Oxford University Press.

———. 1748/2000. *An Enquiry concerning Human Understanding.* Edited by Tom L. Beauchamp. Oxford: Oxford University Press.

———. 1777/1985. *Essays, Moral, Political and Literary.* Edited by Eugene F. Miller. Rev. ed. Indianapolis: Liberty Classics. Based on the edition originally published as vol. 1 of *Essays and Treatises on Several Subjects.*

———. 1778/1983. *The History of England.* 6 vols. 6th ed. Reprint, Indianapolis: Liberty Fund. The 1st edition appeared in 1754–62.

———. 1779/1947. *Dialogues concerning Natural Religion.* Edited by Norman Kemp Smith. New York: Macmillan.

Hundert, E. G. 1994. *The Enlightenment's Fable: Bernard Mandeville and the Discovery of Society.* Cambridge: Cambridge University Press.

Hutchison, Terence. 1978. *On Revolutions and Progress in Economic Knowledge.* Cambridge: Cambridge University Press.

———. 1988. *Before Adam Smith.* Oxford: Basil Blackwell.

Hutton, James. 1788. "Theory of the Earth; or, An Investigation of the Laws Observable in the Composition, Dissolution, and Restoration of Land upon the Globe." *Transactions of the Royal Society of Edinburgh* 1:209–304.

———. 1795. *Theory of the Earth, with Proofs and Illustrations.* London: Cadell, Junior & Davies.

Ingrao, Bruna, and Giorgio Israel. 1990. *The Invisible Hand: Economic Equilibrium in the History of Science.* Translated by Ian McGilvray. Cambridge, MA: MIT Press.

Jackson, Myles W. 1994. "Natural and Artificial Budgets: Accounting for Goethe's Economy of Nature." *Science in Context* 7:409–31.

Jacob, Margaret. 1986. "Christianity and the Newtonian Worldview." In *God and Nature: Historical Essays on the Encounter between Christianity and Science,* ed. David C. Lindberg and Ronald L. Numbers. Berkeley and Los Angeles: University of California Press.

Jacyna, L. S. 1981. "The Physiology of Mind, the Unity of Nature, and the Moral Order in Victorian Thought." *British Journal for the History of Science* 14:109–32.

Jardine, N., J. A. Secord, and E. C. Spary, eds. 1996. *Cultures of Natural History.* Cambridge: Cambridge University Press.

Jennings, Richard. 1855/1969. *Natural Elements of Political Economy.* Reprint, New York: August M. Kelley.

Jevons, William Stanley. 1871/1957. *The Theory of Political Economy.* 5th ed. Reprint, New York: Augustus M. Kelley.

———. 1876/1905. "The Future of Political Economy." Reprinted in *The Principles of Economics and Other Papers,* ed. Henry Higgs. London: Macmillan.

———. 1877. *The Principles of Science.* 2nd ed. London: Macmillan.

———. 1890. "Utilitarianism." In *Pure Logic and Other Minor Works,* ed. Robert Adamson and Harriet A. Jevons. London: Macmillan.

———. 1905. *The Principles of Economics and Other Papers.* Edited by Henry Higgs. London: Macmillan.

———. 1906/1965. *The Coal Question: An Inquiry concerning the Progress of the Nation, and the Probable Exhaustion of Our Coal-Mines.* 3rd rev. ed. Edited by A. W. Flux. Reprint, New York: Augustus M. Kelley. The 1st edition appeared in 1865.

———. 1977. *Papers and Correspondence of William Stanley Jevons.* Vol. 4, *Correspondence, 1873–1878,* ed. R. D. C. Black. London: Macmillan.

Jolink, Albert. 1996. *The Evolutionist Economics of Léon Walras.* London: Routledge.

Kadish, Alon. 1982. *The Oxford Economists in the Late Nineteenth Century.* Oxford: Clarendon.

Kamstra, Mark J., Lisa A. Kramer, and Maurice D. Levi. 2003. "Winter Blues: A SAD Stock Market Cycle." *American Economic Review* 93:324–43.

Kaye, Joel. 1998. *Economy and Nature in the Fourteenth Century.* Cambridge: Cambridge University Press.

Keohane, Nannerl. 1980. *Philosophy and the State in France, the Renaissance to the Enlightenment.* Princeton, NJ: Princeton University Press.

Keynes, John Maynard. 1936/1973. *The General Theory of Employment, Interest and Money,* ed. Donald E. Moggridge. Vol. 7 of *The Collected Writings of John Maynard Keynes,* ed. Donald E. Moggridge and Elizabeth Johnson. New York: Cambridge University Press.

Kindleberger, C. P. 1976. "The Historical Background: Adam Smith and the Industrial Revolution." In *The Market and the State: Essays in Honour of Adam Smith,* ed. Thomas Wilson and A. S. Skinner. Oxford: Clarendon.

Kingsland, Sharon. 1994. "Economics and Evolution: Alfred James Lotka and the Economy of Nature." In *Natural Images in Economic Thought,* ed. Philip Mirowski. Cambridge: Cambridge University Press.

Klant, J. J. 1988. "The Natural Order." In *The Popperian Legacy in Economics,* ed. Neil De Marchi. Cambridge: Cambridge University Press.

Kleer, Richard. 1993. "Adam Smith on the Morality of the Pursuit of Fortune." *Economics and Philosophy* 9:289–95.

———. 1995. "Final Causes in Adam Smith's *Theory of Moral Sentiments.*" *Journal of the History of Philosophy* 33:275–300.

Klein, Daniel. 1985. "Deductive Economic Methodology in the French Enlightenment: Condillac and Destutt de Tracy." *History of Political Economy* 17:51–71.

Knoepflmacher, U. C., and G. B. Tennyson, eds. 1977. *Nature and the Victorian Imagination.* Berkeley and Los Angeles: University of California Press.

Koerner, Lisbet. 1999. *Linnaeus: Nature and Nation.* Cambridge, MA: Harvard University Press.

Kohn, David, ed. 1985. *The Darwinian Heritage.* Princeton, NJ: Princeton University Press.

Kramer, Charles. 1994. "Macroeconomic Seasonality and the January Effect." *Journal of Finance* 49:1883–91.

Kuhn, Thomas S. 1957. *The Copernican Revolution.* Cambridge, MA: Harvard University Press.

———. 1970. *The Structure of Scientific Revolutions.* 2nd ed. Chicago: University of Chicago Press.

———. 1977. *The Essential Tension.* Chicago: University of Chicago Press.

Lagueux, Maurice. 1999. "Do Metaphors Affect Economic Theory?" *Economics and Philosophy* 15 (1): 1–22.

Laidler, David. 1991. "The Quantity Theory Is Always and Everywhere Controversial—Why?" *Economic Record* 67:289–306.

Lallement, Jérôme. 2000. "Prix et équilibre selon Léon Walras." In *Nouvelle histoire de la pensée économique,* ed. Alain Béraud and Gilbert Faccarello, vol. 2. Paris: Editions la Découverte.

La Mettrie, Julien Offray. 1758/1912. *Man a Machine.* Translated by Gertrude C. Bussey. Revised by M. W. Calkins. Chicago: Open Court Classics.

Larrère, Catherine. 1992. *L'invention de l'économie au XVIIIe siècle.* Paris: Presses Universitaires de France.

Latour, Bruno. 1993. *We Have Never Been Modern.* Translated by Catherine Porter. Cambridge, MA: Harvard University Press.

Laudan, Rachel. 1987. *From Mineralogy to Geology.* Chicago: University of Chicago Press.

Lauderdale, James Maitland. 1804. *Inquiry into the Nature and Origin of Public Wealth.* London: Longman & Rees.

Lavoisier, A.-L. 1893/1965. "Éloge de M. de Colbert." In *Oeuvres de Lavoisier,* vol. 6, *Rapports à l'Académie, notes et rapports divers. . . .* Reprint, New York: Johnson Reprint. The "Éloge" was first published in 1771.

Lawrence, Christopher. 1979. "The Nervous System and Society in the Scottish Enlightenment." In *Natural Order: Historical Studies of Scientific Culture,* ed. Barry Barnes and Steven Shapin. Beverly Hills, CA: Sage.

———. 1982. "Joseph Black: The Natural Philosophical Background." In *Joseph Black, 1728–1799: A Commemorative Symposium,* ed. A. D. C. Simpson. Edinburgh: Royal Scottish Museum.

Leary, David E. 1983. "The Fate and Influence of John Stuart Mill's Proposed Science of Ethology." *Journal of the History of Ideas* 43:153–62.

Lennox, James G. 2001. *Aristotle's Philosophy of Biology.* Cambridge: Cambridge University Press.

Lenzer, Gertrud, ed. 1975. *Auguste Comte and Positivism: The Essential Writings.* Chicago: University of Chicago Press.

Lepenies, Wolf. 1982. "Linnaeus's *Nemesis divina* and the Concept of Divine Retaliation." *Isis* 73:11–27.

Levere, Trevor. 1981. *Poetry Realized in Nature.* Cambridge: Cambridge University Press.

Levine, A. L. 1983. "Marshall's *Principles* and the 'Biological Viewpoint': A Reconsideration." *Manchester School* 51:276–93.

Levy, David M. 1982. "Rational Choice and Morality: Economics and Classical Philosophy." *History of Political Economy* 14:1–36.

———. 1992. *The Economic Ideas of Ordinary People.* London: Routledge.

———. 2001. *How the Dismal Science Got Its Name: Classical Economics and the Ur-Text of Racial Politics.* Ann Arbor: University of Michigan Press.

Liedman, Sven-Eric. 1989. "Utilitarianism and the Economy." In *Science in Sweden: The Royal Swedish Academy of Sciences, 1739–1989,* ed. Tore Frängsmyr. Canton, MA: Science History.

Limoges, Camille. 1970. *La selection naturelle: Étude sur la première constitution d'un concept.* Paris: Presses Universitaires de France.

Limoges, Camille, and Claude Ménard. 1994. "Organization and the Division of Labor: Biological Metaphors at Work in Alfred Marshall's *Principles of Economics.*" In *Natural Images in Economic Thought,* ed. Philip Mirowski. Cambridge: Cambridge University Press.

Lindberg, David C., and Ronald L. Numbers, eds. 1986. *God and Nature: Historical Essays on the Encounter between Christianity and Science.* Berkeley and Los Angeles: University of California Press.

Linnaeus, Carl. 1791/1977a. *Oeconomy of Nature.* In *Miscellaneous Tracts Relating to Natural History, Husbandry, and Physick* (3rd ed.), ed. and trans. Benjamin Stillingfleet. Reprint, New York: Arno. The *Oeconomia naturae* first appeared in 1749 as an Uppsala University doctoral thesis (purportedly by Isaac J. Biberg) and was first published in 1750 in Stockholm by Kieswetters boklådor.

———. 1791/1977b. *Polity of Nature.* In *Miscellaneous Tracts Relating to Natural History, Husbandry, and Physick* (3rd ed.), ed. and trans. Benjamin Stillingfleet. Reprint, New York: Arno. The *Politia naturae* first appeared in 1760.

———. 1968. *Nemesis divina.* Edited by Elis Malmeström and Telemak Fredbärj. Stockholm: Bonniers.

Livingston, Donald W. 1984. *Hume's Philosophy of Common Life.* Chicago: University of Chicago Press.

Lloyd, G. E. R. 1970. *Early Greek Science: Thales to Aristotle.* New York: Norton.

———. 1987. *The Revolutions of Wisdom: Studies in the Claims and Practice of Ancient Greek Science.* Berkeley and Los Angeles: University of California Press.

———. 1992. "Greek Antiquity." In *The Concept of Nature,* ed. John Torrance. Oxford: Clarendon.

Locke, John. 1696/1989. *Several Papers Relating to Money, Interest and Trade, &c.* Reprint, New York: Augustus M. Kelley.

———. 1764/1980. *Second Treatise of Government.* 6th ed. Edited, with an introduction, by C. B. Macpherson. Indianapolis: Hackett. The 1st edition was published in 1690.

Lovejoy, Arthur O. 1936. *The Great Chain of Being.* Cambridge, MA: Harvard University Press.

Lyell, Charles. 1830–33. *Principles of Geology.* 3 vols. London: John Murray. A facsimile reprint, with an introduction by Martin J. S. Rudwick, was issued by the University of Chicago Press in 1990.

Mably, Gabriel Bonnot de. 1768. *Doutes proposés aux philosophes économists dur l'ordre naturel et essential des sociétés politiques.* Paris: A la Haye. I used the copy at Harvard University's Kress Library.

Macfie, A. L. 1961. "Adam Smith's Theory of Moral Sentiments." *Scottish Journal of Political Economy* 8:12–27.

Macpherson, C. B. 1962. *The Political Theory of Possessive Individualism: Hobbes to Locke.* Oxford: Oxford University Press.

Maloney, John. 1985. *Marshall, Orthodoxy, and the Professionalisation of Economics.* Cambridge: Cambridge University Press.

Malthus, Thomas Robert. 1803/1989. *An Essay on the Principle of Population.* 2 vols. Edited by Patricia James. Cambridge: Cambridge University Press. The 1st edition appeared in 1798. James's edition is based on the 1803 enlarged 2nd edition and includes the variora of 1806, 1807, 1817, and 1826.

———. 1820/1989. *Principles of Political Economy.* 2 vols. Edited by John Pullen. Cambridge: Cambridge University Press. Pullen's edition, which reprints the 1820 1st edition, includes in vol. 2 a list of the alterations to the 1st edition found in Malthus's own "Manuscript Revisions" and in the posthumous 1832 2nd edition.

———. 1824/1986. "On Political Economy." In *Works of Thomas Robert Malthus,* vol. 7, *Essays on Political Economy.* London: William Pickering.

Mandeville, Bernard. 1970. *The Fable of the Bees.* Edited by Phillip Harth. London: Penguin.

Marshall, Alfred. 1867. "The Law of Parcimony." Cambridge University, Marshall Library of Economics, Alfred Marshall Papers, box 11 (11). This essay was first published in Tiziano Raffaelli, ed., "The Early Philosophical Writings of Alfred Marshall," *Research in the History of Economic Thought and Methodology* 4, archival suppl. (1994): 53–159.

———. 1890/1920. *The Principles of Economics.* 8th ed. London: Macmillan.

———. 1898/1925. "Mechanical and Biological Analogies in Economics." Reprinted in *The Memorials of Alfred Marshall,* ed. Alfred C. Pigou. London: Macmillan.

Marx, Leo. 1964. *The Machine in the Garden: Technology and the Pastoral Ideal in America.* Oxford: Oxford University Press.

McCulloch, John R. 1825. *Outlines of Political Economy.* Edited by Rev. John McVickar. New York: Wilder & Campbell.

———. 1853/1967. *Treatises and Essays on Subjects Connected with Economic Policy.* Reprint, New York: Augustus M. Kelley.

———. 1864/1965. *The Principles of Political Economy.* 5th ed. Reprint, New York: Augustus M. Kelley. The 1st edition appeared in 1825.

McKendrick, Neil, John Brewer, and J. H. Plumb. 1982. *The Birth of a Consumer Society: The Commercialization of Eighteenth-Century England.* Bloomington: Indiana University Press.

McNally, David. 1988. *Political Economy and the Rise of Capitalism: A Reinterpretation.* Berkeley and Los Angeles: University of California Press.

McWilliams, Rita. 1969. "The Papers of Alfred Marshall Deposited in the Marshall Library." *History of Economic Thought Newsletter* 3:9–19.

Meek, Ronald. 1962. *The Economics of Physiocracy: Essays and Translations.* London: George Allen & Unwin.

———. 1971. "Smith, Turgot, and the 'Four Stages' Theory." *History of Political Economy* 3:9–27.

———. 1973. *Turgot on Progress, Sociology and Economics.* Cambridge: Cambridge University Press.

Megill, A. D. 1975. "Theory and Experience in Adam Smith." *Journal of the History of Ideas* 36:79–94.

Meikle, Scott. 1991. "History of Philosophy: The Metaphysics of Substance in Marx." In *The Cambridge Companion to Marx,* ed. Terrell Carver. Cambridge: Cambridge University Press.

Melon, Jean François. 1734/1739. *A Political Essay upon Commerce.* Translated by David Bindon. Dublin. The original French text can be found in Eugene Daire, ed., *Economistes financières du dix-huitième siècle* (1843; Geneva: Slatkine, 1971), 665–778.

Ménard, Claude. 1978. *La formulation d'une rationalité économique: A. A. Cournot.* Paris: Flammarion.

Mercier de la Rivière, Pierre. 1767/1910. *L'order naturel et essentiel des sociétés politiques.* Reprint, Paris: Paul Geuthner.

Merton, Robert K. 1938. "Science, Technology and Society in Seventeenth Century England." *Osiris* 4:360–632.

Milgate, Murray, and Shannon C. Stimson. 1991. *Ricardian Politics.* Princeton, NJ: Princeton University Press.

Mill, John Stuart. 1836/1967. "On the Definition of Political Economy and on the Method of Investigation Proper to It." In *Collected Works of John Stuart Mill,* vol. 4, *Essays on Economics and Society.* Toronto: University of Toronto Press.

———. 1859/1978. "Bain's Psychology." In *Collected Works of John Stuart Mill,* vol. 11, *Essays on Philosophy and the Classics.* Toronto: University of Toronto Press.

———. 1871/1965. *Principles of Political Economy.* 7th ed. Vols. 2–3 of *Collected Works of John Stuart Mill.* Toronto: University of Toronto Press. The 1st edition appeared in 1848.

———. 1872/1973. *A System of Logic: Ratiocinative and Inductive.* 8th ed. Vols. 7–8 of *Collected Works of John Stuart Mill.* Toronto: University of Toronto Press. The 1st edition appeared in 1843.

———. 1874/1969. "On Nature." In *Collected Works of John Stuart Mill,* vol. 10, *Essays on Ethics, Religion, and Society.* Toronto: University of Toronto Press.

———. 1963–91. *Collected Works of John Stuart Mill,* ed. John M. Robson. 33 vols. Toronto: University of Toronto Press.

———. 1967. *Essays on Economics and Society.* Vols. 4–5 of *Collected Works of John Stuart Mill.* Toronto: University of Toronto Press.

———. 1969. *Essays on Ethics, Religion, and Society.* Vol. 10 of *Collected Works of John Stuart Mill.* Toronto: University of Toronto Press.

———. 1972. *The Later Letters of John Stuart Mill, 1849–1873.* Vol. 15 of *Collected Works of John Stuart Mill.* Toronto: University of Toronto Press.

———. 1978. *Essays on Philosophy and the Classics.* Vol. 11 of *Collected Works of John Stuart Mill.* Toronto: University of Toronto Press.

"Mini-Symposium on Physiocracy." 2002. *Journal of the History of Economic Thought* 24:39–110.

Minowitz, Peter. 1993. *Profits, Priests, and Princes: Adam Smith's Emancipation of Economics from Politics and Religion.* Stanford, CA: Stanford University Press.

Mirabeau, Victor. 1763. *Philosophie rurale; ou, Économie générale et politique de l'agriculture.* Amsterdam: Chez les Libraires associés.

Mirowski, Philip. 1987. "Shall I Compare Thee to a Minkowski-Ricardo-Leontief-Metzler Matrix of the Mosak-Hicks Type? or, Rhetoric, Mathematics, and the Nature of Neoclassical Economic Theory." *Economics and Philosophy* 3:67–95.

———. 1989. *More Heat Than Light.* Cambridge: Cambridge University Press.

———. 1994a. "Marshalling the Unruly Atoms: Understanding Edgeworth's Career." In *Edgeworth on Chance, Economic Hazard, and Statistics,* ed. Philip Mirowski. Lanham, MD: Rowman & Littlefield.

———, ed. 1994b. *Natural Images in Economic Thought.* Cambridge: Cambridge University Press.

———. 2002. *Machine Dreams: Economics Becomes a Cyborg Science.* Cambridge: Cambridge University Press.

Mischel, Theodore. 1965. "'Emotion' and 'Motivation' in the Development of English Psychology: D. Hartley, James Mill, A. Bain." *Journal of the History of the Behavioral Sciences* 1:123–44.

Mitchell, Neil J. 1986. "John Locke and the Rise of Capitalism." *History of Political Economy* 18:291–305.

Mizuta, Hiroshi. 1967/2000. *Adam Smith's Library.* 2nd ed. Cambridge: Cambridge University Press.

Mokyr, Joel. 2002. *The Gifts of Athena: Historical Origins of the Knowledge Economy.* Princeton, NJ: Princeton University Press.

Monchrétien, Antoyne de. 1615/1939. *Traicté de l'oeconomie politique.* Reprint, Paris: Marcel Rivière.

Montes, Leonidas. 2004. *Adam Smith in Context.* Basingstoke: Palgrave Macmillan.

Montesquieu, Charles Louis. 1748/1989. *The Spirit of the Laws.* Translated by Anne M. Cohler, Basia Carolyn Miller, and Harold Samuel Stone. Cambridge: Cambridge University Press.

Moore, James. 1979. "The Social Background of Hume's Science of Human Nature." In *McGill Hume Studies,* ed. D. F. Norton, N. Capaldi, and W. L. Robison. San Diego: Austin Hill.

Moore, James, and Michael Silverthorne. 1983. "Gershom Carmichael and the Natural Jurisprudence Tradition in Eighteenth-Century Scotland." In *Wealth and Virtue,* ed. Istvan Hont and Michael Ignatieff. Cambridge: Cambridge University Press.

Moore, James R. 1986. "Geologists and Interpreters of Genesis in the Nineteenth Century." In *God and Nature: Historical Essays on the Encounter between Christianity and Science,* ed. David C. Lindberg and Ronald L. Numbers. Berkeley and Los Angeles: University of California Press.

Morgan, Mary S. 1990. *The History of Econometric Ideas.* Cambridge: Cambridge University Press.

Morgan, Mary S., and Margaret Morrison, eds. 1999. *Models as Mediators.* Cambridge: Cambridge University Press.

Mosca, Manuela. 1998. "Jule Dupuit, the French 'Ingénieurs Économistes' and the Société d'Économie Politique." In *Studies in the History of French Political Economy: From Bodin to Walras,* ed. Gilbert Faccarello. New York: Routledge.

Moss, Laurence. 1982. "Biological Theory and Technological Entrepreneurship in Marshall's Writings." *Eastern Economic Journal* 18:3–13.

Mossner, Ernest Campbell. 1980. *The Life of David Hume.* 2nd ed. Oxford: Clarendon.

Mukerji, Chandra. 1993. "Reading and Writing with Nature: A Materialist Approach to French Formal Gardens." In *Consumption and the World of Goods,* ed. John Brewer and Roy Porter. London: Routledge.

Müller-Wille, Staffan. 2003. "Nature as a Marketplace: The Political Economy of Linnaean Botany." In *Oeconomies in the Age of Newton,* ed. Margaret Schabas and Neil De Marchi. Durham, NC: Duke University Press.

Murphy, Antoin E. 1986. *Cantillon: Entrepreneur and Economist.* Oxford: Oxford University Press.

———. 1997. *John Law: Economic Theorist and Policy-Maker.* Oxford: Clarendon.

Newton, Isaac. 1687/1934. *Principia.* Translated by Florian Cajori. Berkeley: University of California Press.

———. 1704/1979. *Opticks*. Reprint, New York: Dover.

Niman, Neil B. 1991. "Biological Analogies in Marshall's Work." *Journal of the History of Economic Thought* 13:19–36.

Noble, David. 1997. *The Religion of Technology*. New York: Alfred A. Knopf.

Norton, D. F., N. Capaldi, and W. L. Robison, eds. 1979. *McGill Hume Studies*. San Diego: Austin Hill.

Numbers, Ronald L. 1986. "The Creationists." In *God and Nature: Historical Essays on the Encounter between Christianity and Science*, ed. David C. Lindberg and Ronald L. Numbers. Berkeley and Los Angeles: University of California Press.

O'Brien, Denis P. 1970. *J. R. McCulloch: A Study in Classical Economics*. London: George Allen & Unwin.

———. 1975. *The Classical Economists*. Oxford: Oxford University Press.

Olson, Richard. 1975. *Scottish Philosophy and British Physics, 1750–1880*. Princeton, NJ: Princeton University Press.

Orain, Arnaud. 2003. "Decline and Progress: The Economic Agent in Condillac's Theory of History." *European Journal of the History of Economic Thought* 10:379–407.

Oslington, Paul. 2001. "John Henry Newman, Nassau Senior, and the Separation of Political Economy from Theology in the Nineteenth Century." *History of Political Economy* 33:825–42.

Oswald, Donald J. 1995. "Metaphysical Beliefs and the Foundations of Smithian Political Economy." *History of Political Economy* 27:449–76.

Otteson, James R. 2002. *Adam Smith's Marketplace of Life*. Cambridge: Cambridge University Press.

Paradis, James. 1989. "*Evolution and Ethics* in Its Victorian Context." In *Evolution and Ethics,* ed. James Paradis and George C. Williams. Princeton, NJ: Princeton University Press.

Peach, Terry. 1993. *Interpreting Ricardo*. Cambridge: Cambridge University Press.

Peart, Sandra. 1996. *The Economics of W. S. Jevons*. London: Routledge.

Penrose, Roger. 1992. "The Modern Physicist's View of Nature." In *The Concept of Nature,* ed. John Torrance. Oxford: Clarendon.

Perkins, Jean A. 1979. "The Physiocrats and the Encyclopedists." *Studies in Eighteenth-Century Culture* 8:323–36.

Perlman, Morris. 1987. "Of a Controversial Passage in Hume." *Journal of Political Economy* 95:274–89.

Philosophical Society of Edinburgh. 1754–71. *Essays and Observations, Physical and Literary*. 3 vols. Edinburgh: G. Hamilton & J. Balfour. Volumes 1 (1754) and 2 (1756) were edited by David Hume and Alexander Monro. I used the copy in the University of Toronto's Fisher Rare Books Library.

Pigou, Alfred C., ed. 1925. *The Memorials of Alfred Marshall*. London: Macmillan.

Pitson, Antony E. 1993. "The Nature of Humean Animals." *Hume Studies* 19:301–16.

Pluche, Noël-Antoine. 1732–50. *Le spectacle de la nature.* 8 vols. Paris: Estienne & fils.

Poirier, Jean-Pierre. 1993. *Lavoisier: Chemist, Biologist, Economist.* Translated by Rebecca Balinski. Philadelphia: University of Pennsylvania Press.

Polanyi, Karl. 1957. "Aristotle Discovers the Economy." In *Trade and Market in the Early Empires,* ed. Karl Polanyi et al. New York: Free Press.

Poovey, Mary. 1998. *A History of the Modern Fact: Problems of Knowledge in the Sciences of Wealth and Society.* Chicago: University of Chicago Press.

Porter, Roy. 1979. "Creation and Credence: The Career of Theories of the Earth in Britain, 1660–1820." In *Natural Order: Historical Studies of Scientific Culture,* ed. Barry Barnes and Steven Shapin. Beverly Hills, CA: Sage.

———. 1995. "Medical Science and Human Science." In *Inventing Human Science: Eighteenth-Century Domains,* ed. Christopher Fox, Roy Porter, and Robert Wokler. Berkeley and Los Angeles: University of California Press.

———, ed. 2003. *Eighteenth-Century Science.* Vol. 4 of *The Cambridge History of Science.* Cambridge: Cambridge University Press.

Porter, Roy, and Mikuláš Teich, eds. 1981. *The Enlightenment in National Context.* Cambridge: Cambridge University Press.

Porter, Theodore M. 1990. "Natural Science and Social Theory." In *Companion to the History of Modern Science,* ed. R. C. Olby, G. N. Cantor, J. R. R. Christie, and M. J. S. Hodge. London: Routledge.

———. 1995. *Trust in Numbers: The Pursuit of Objectivity in Science and Public Life.* Princeton, NJ: Princeton University Press.

Porter, Theodore M., and Dorothy Ross, eds. 2003. *The Modern Social Sciences.* Vol. 7 of *The Cambridge History of Science.* Cambridge: Cambridge University Press.

Prieto, Jimena Hurtado. 2004. "Bernard Mandeville's Heir: Adam Smith or Jean-Jacques Rousseau on the Possibility of Economic Analysis." *European Journal of the History of Economic Thought* 11:1–31.

Pullen, J. M. 1981. "Malthus' Theological Ideas and Their Influence on His Principle of Population." *History of Political Economy* 13:39–54.

———. 1995. "Malthus on Agricultural Protection: An Alternative View." *History of Political Economy* 27:517–29.

Puro, Edward. 1992. "Use of the Term 'Natural' in Adam Smith's Wealth of Nations." *Research in the History of Economic Thought and Methodology* 9:73–86.

Quesnay, François. 1736. *Essai phisique sur l'oeconomie animale.* Paris: Guillaume Cavelier.

———. 1758–67/1962. "Tableau économique." In *The Economics of Physiocracy: Essays and Translations,* ed. Ronald Meek. London: George Allen & Unwin.

———. 1765/1962. "Natural Right." In *The Economics of Physiocracy: Essays and Translations,* ed. Ronald Meek. London: George Allen & Unwin.

———. 1766/1962. "Dialogue on the Work of Artisans." In *The Economics of*

Physiocracy: Essays and Translations, ed. Ronald Meek. London: George Allen & Unwin.

Raffaelli, Tiziano. 1991. "The Analysis of the Human Mind in the Early Marshallian Manuscripts." *Quaderni di Storia dell'Economia Politica* 9:29–58.

———, ed. 1994. "The Early Philosophical Writings of Alfred Marshall." *Research in the History of Economic Thought and Methodology* 4 (archival suppl.): 53–159.

Rashid, Salim. 1981a. "Malthus' *Principles* and British Economic Thought, 1820–1835." *History of Political Economy* 13:55–79.

———. 1981b. "Political Economy and Geology in the Early Nineteenth Century: Similarities and Contrasts." *History of Political Economy* 13:726–44.

———. 1984. "David Hume and Eighteenth Century Monetary Thought: A Critical Comment on Recent Views." *Hume Studies* 10:154–64.

Rausing, Lisbet. 2003. "Underwriting the Oeconomy: Linnaeus on Nature and Mind." In *Oeconomies in the Age of Newton,* ed. Margaret Schabas and Neil De Marchi. Durham, NC: Duke University Press.

Ray, John. 1692. *The Wisdom of God Manifested in the Work of the Creation.* 2nd ed. London: Samuel Smith. The 1st edition appeared in 1691.

Redman, Deborah A. 1997. *The Rise of Political Economy as a Science.* Cambridge, MA: MIT Press.

Reid, Gavin C. 1972. "Jevons's Treatment of Dimensionality in *The Theory of Political Economy:* An Essay in the History of Mathematical Economics." *Manchester School* 40:85–98.

Reill, Peter. 2003. "The Legacy of the 'Scientific Revolution': Science and the Enlightenment." In *Eighteenth-Century Science (The Cambridge History of Science,* vol. 4), ed. Roy Porter. Cambridge: Cambridge University Press.

Ricardo, David. 1817/1951. *On the Principles of Political Economy and Taxation.* Vol. 1 of *Works and Correspondence,* ed. Piero Sraffa. Cambridge: Cambridge University Press.

———. 1951–73. *Works and Correspondence.* Edited by Piero Sraffa. 9 vols. Cambridge: Cambridge University Press.

———. 1962. *Letters, 1810–1815.* Vol. 6 of *Works and Correspondence,* ed. Piero Sraffa. Cambridge: Cambridge University Press.

Richards, Robert J. 1987. *Darwin and the Emergence of Evolutionary Theories of Mind and Behavior.* Chicago: University of Chicago Press.

———. 2002. *The Romantic Conception of Life: Science and Philosophy in the Age of Goethe.* Chicago: University of Chicago Press.

Riskin, Jessica. 1998. "Poor Richard's Leyden Jar: Electricity and Economy in Franklinist France." *Historical Studies in the Physical and Biological Sciences* 28:301–36.

———. 2002. *Science in the Age of Sensibility.* Chicago: University of Chicago Press.

———. 2003. "The 'Spirit of System' and the Fortunes of Physiocracy." In *Oecon-*

omies in the Age of Newton, ed. Margaret Schabas and Neil De Marchi. Durham, NC: Duke University Press.

Robinson, Bryan. 1734/1737. *A Treatise of the Animal Oeconomy.* 2nd ed. London: S. Powell.

Robson, John M. 1976. "Rational Animals and Others." In *James and John Stuart Mill Papers of the Centenary Conference,* ed. John M. Robson and Michael Laine. Toronto: University of Toronto Press.

Roe, Shirley. 1985. "Voltaire versus Needham: Atheism, Materialism, and the Generation of Life." *Journal of the History of Ideas* 46:65–87.

Romano, Richard M. 1982. "The Economic Ideas of Charles Babbage." *History of Political Economy* 14:385–405.

Rorty, Amélie. 1982. "From Passions to Emotions and Sentiments." *Philosophy* 57:159–72.

Rosenberg, Alexander. 1976. *Microeconomic Laws.* Pittsburgh: University of Pittsburgh Press.

Rosenberg, Nathan. 1990. "Adam Smith and the Stock of Moral Capital." *History of Political Economy* 22:1–17.

Ross, Ian Simpson. 1995. *The Life of Adam Smith.* Oxford: Clarendon.

Rothschild, Emma. 1992. "Adam Smith and Conservative Economics." *Economic History Review* 45:74–96.

———. 1994. "Adam Smith and the Invisible Hand." *American Economic Review* 84:319–22.

———. 2001. *Economic Sentiments: Adam Smith, Condorcet, and the Enlightenment.* Cambridge, MA: Harvard University Press.

Rotwein, Eugene. 1970. Introduction to *David Hume: Writings on Economics,* ed. Eugene Rotwein. Madison: University of Wisconsin Press.

Rousseau, G. S., and Roy Porter, eds. 1980. *The Ferment of Knowledge: Studies in the Historiography of Eighteenth-Century Science.* Cambridge: Cambridge University Press.

Rousseau, Jean-Jacques. 1781–82/2000. *The Reveries of the Solitary Walker; Botanical Writings; and Letter to Franquières.* Edited by Christopher Kelly. Vol. 8 of *The Collected Writings of Jean-Jacques Rousseau.* Hanover, NH: University Press of New England.

———. 1987. *Basic Political Writings.* Translated and edited by Donald A. Cress. Indianapolis: Hackett.

Rudwick, Martin J. S. 1971. "Uniformity and Progression: Reflections on the Structure of Geological Theory in the Age of Lyell." In *Perspectives in the History of Science and Technology,* ed. Duane H. D. Roller. Norman: University of Oklahoma Press.

———. 1974. "Poulett Scrope on the Volcanoes of Auvergne: Lyellian Time and Political Economy." *British Journal for the History of Science* 7:205–42.

———. 1976. *The Meaning of Fossils: Episodes in the History of Palaeontology.* Chicago: University of Chicago Press.

———. 1979. "Transposed Concepts from the Human Species in the Early Work of Charles Lyell." In *Images of the Earth: Essays in the History of the Environmental Sciences,* ed. L. S. Jordanova and Roy Porter. Chalfort St. Giles: British Society for the History of Science.

Ruestow, Edward G. 1973. *Physics at Seventeenth and Eighteenth-Century Leiden: Philosophy and the New Science in the University.* The Hague: Martinus Nijhoff.

Runde, Jochen, and Sohei Mizuhara, eds. 2003. *The Philosophy of Keynes's Economics: Probability, Uncertainty and Convention.* London: Routledge.

Ruse, Michel. 1979. *The Darwinian Revolution.* Chicago: University of Chicago Press.

Russell, Paul. 1995. *Freedom and Moral Sentiment: Hume's Way of Naturalizing Responsibility.* Oxford: Oxford University Press.

Rutherford, Malcolm. 1996. *Institutions in Economics: The Old and the New Institutionalism.* Cambridge: Cambridge University Press.

Ryan, Alan. 1974. *J. S. Mill.* London: Routledge & Kegan Paul.

———. 1990. *The Philosophy of John Stuart Mill.* 2nd ed. Atlantic Highlands, NJ: Humanities.

Ryazanskaya, S. W., ed. 1955/1975. *Marx-Engels Selected Correspondence.* Translated by I. Lasker. 3rd ed. Moscow: Progress.

Samuels, Warren. 1966. *The Classical Theory of Economic Policy.* New York: World.

Sapadin, Eugene. 1997. "A Note on Newton, Boyle, and Hume's 'Experimental Method.'" *Hume Studies* 23:337–44.

Schabas, Margaret. 1990a. "Ricardo Naturalized: Lyell and Darwin on the Economy of Nature." In *Perspectives on the History of Economic Thought,* vol. 3, ed. Donald E. Moggridge. Aldershot: Edward Elgar.

———. 1990b. *A World Ruled by Number: William Stanley Jevons and the Rise of Mathematical Economics.* Princeton, NJ: Princeton University Press.

———. 1992. "Breaking Away: History of Economics as History of Science." *History of Political Economy* 24:187–203.

———. 1994a. "The Greyhound and the Mastiff: Darwinian Themes in Mill and Marshall." In *Natural Images in Economic Thought,* ed. Philip Mirowski. Cambridge: Cambridge University Press.

———. 1994b. "Market Contracts in the Age of Hume." In *Higgling: Transactors and Their Markets in the History of Economics,* ed. Neil De Marchi and Mary S. Morgan. Durham, NC: Duke University Press.

———. 1995a. "John Stuart Mill and Concepts of Nature." *Dialogue* 34:447–65.

———. 1995b. "Parmenides and the Cliometricians." In *The Reliability of Economic Models: Essays in the Epistemology of Economics,* ed. Daniel C. Little. Dordrecht: Kluwer.

———. 1997. "Victorian Economics and the Science of the Mind." In *Victorian Science in Context,* ed. Bernard Lightman. Chicago: University of Chicago Press.

———. 2001. "David Hume on Experimental Natural Philosophy, Money, and Fluids." *History of Political Economy* 33:411–35.

———. 2002. "Coming Together: History of Economics as History of Science." In *The Future of the History of Economics,* ed. E. Roy Weintraub. Durham, NC: Duke University Press.

———. 2003. "Smith's Debts to Nature." In *Oeconomies in the Age of Newton,* ed. Margaret Schabas and Neil De Marchi. Durham, NC: Duke University Press.

Schabas, Margaret, and Neil De Marchi, eds. 2003. *Oeconomies in the Age of Newton.* Durham, NC: Duke University Press.

Schaffer, Simon. 1983. "Natural Philosophy and Public Spectacle in the Eighteenth Century." *History of Science* 21:1–43.

———. 1989. "Defoe's Natural Philosophy and the Worlds of Credit." In *Nature Transfigured: Science and Literature, 1700–1900,* ed. J. Christie and S. Shuttleworth. Manchester: Manchester University Press.

———. 1990. "States of Mind: Enlightenment and Natural Philosophy." In *The Languages of Psyche: Mind and Body in Enlightenment Thought,* ed. G. S. Rousseau. Berkeley and Los Angeles: University of California Press.

———. 1993. "The Consuming Flame: Electrical Showmen and Tory Mystics in the World of Goods." In *Consumption and the World of Goods,* ed. John Brewer and Roy Porter. London: Routledge.

———. 1997. "The Earth's Fertility as a Social Fact in Early Modern Britain." In *Nature and Society in Historical Context,* ed. Mikuláš Teich, Roy Porter, and Bo Gustafsson. Cambridge: Cambridge University Press.

———. 1999. "Enlightened Automata." In *The Sciences in Enlightened Europe,* ed. William Clark, Jan Golinski, and Simon Schaffer. Chicago: University of Chicago Press.

Schiebinger, Londa. 1993. *Nature's Body: Gender in the Making of Modern Science.* Boston: Beacon.

Schneewind, J. B., ed. 1970. *Mill.* London: Macmillan.

Schofield, Robert E. 1970. *Mechanism and Materalism: British Natural Philosophy in an Age of Reason.* Princeton, NJ: Princeton University Press.

Schumpeter, Joseph A. 1954. *A History of Economic Analysis.* New York: Oxford University Press.

Schweber, S. S. 1977. "Darwin and the Political Economists: Divergence of Character." *Journal of the History of Biology* 10:195–289.

———. 1985. "The Wider British Context in Darwin's Theorizing." In *The Darwinian Heritage,* ed. David Kohn. Princeton, NJ: Princeton University Press.

Scrope, John Poulett. 1833/1969. *Principles of Political Economy.* Reprint, New York: Augustus M. Kelley.

Searle, John R. 1995. *The Construction of Social Reality.* New York: Free Press.

Seidler, Michael. 1977. "Hume and the Animals." *Southern Journal of Philosophy* 15:361–72.

Senior, Nassau W. 1836/1965. *An Outline of the Science of Political Economy.* Reprint, New York: Augustus M. Kelley.

Shapin, Steven, and Simon Schaffer. 1985. *Leviathan and the Air-Pump: Hobbes, Boyle, and the Experimental Life.* Princeton, NJ: Princeton University Press.

Simpson, A. D. C., ed. 1982. *Joseph Black, 1728–1799: A Commemorative Symposium.* Edinburgh: Royal Scottish Museum.

Skinner, Andrew S. 1967. "Natural History in the Age of Adam Smith." *Political Studies* 15:32–48.

———. 1974. "Adam Smith, Science and the Role of the Imagination." In *Hume and the Enlightenment,* ed. William B. Todd. Edinburgh: University of Edinburgh Press.

———. 1990. "The Shaping of Political Economy in the Enlightenment." *Scottish Journal of Political Economy* 37:145–64.

———. 1993a. "Adam Smith: The Origins of the Exchange Economy." *European Journal of the History of Economic Thought* 1:21–46.

———. 1993b. "David Hume: Principles of Political Economy." In *The Cambridge Companion to Hume,* ed. David Fate Norton. Cambridge: Cambridge University Press.

Skinner, Andrew, and T. Wilson, eds. 1975. *Essays on Adam Smith.* Oxford: Clarendon.

Sloan, Phillip. 1995. "The Gaze of Natural History." In *Inventing Human Science: Eighteenth-Century Domains,* ed. Christopher Fox, Roy Porter, and Robert Wokler. Berkeley and Los Angeles: University of California Press.

Smith, Adam. 1776/1976. *An Inquiry into the Nature and Causes of the Wealth of Nations.* Edited by R. H. Campbell and A. S. Skinner. 2 vols. Oxford: Clarendon.

———. 1790/1976. *The Theory of Moral Sentiments.* 6th ed. Edited by D. D. Raphael and A. L. Macfie. Oxford: Clarendon. The 1st edition appeared in 1759.

———. 1795/1980. *Essays on Philosophical Subjects.* Edited by W. P. D. Wightman. Oxford: Clarendon.

———. 1977. *Correspondence of Adam Smith.* Edited by E. C. Mossner and I. S. Ross. Oxford: Clarendon.

———. 1978. *Lectures on Jurisprudence.* Edited by R. L. Meek, D. D. Raphael, and P. G. Stein. Oxford: Oxford University Press. The lectures date to 1762–63 and 1766.

Smith, Crosbie. 1998. *The Science of Energy: A Cultural History of Energy Physics in Victorian Britain.* Chicago: University of Chicago Press.

Smith, Pamela H. 1994. *The Business of Alchemy: Science and Culture in the Holy Roman Empire.* Princeton, NJ: Princeton University Press.

Smith, Roger. 1973. "The Background of Physiological Psychology in Natural Philosophy." *History of Science* 11:75–123.

———. 1995. "The Language of Human Nature." In *Inventing Human Science: Eighteenth-Century Domains,* ed. Christopher Fox, Roy Porter, and Robert Wokler. Berkeley and Los Angeles: University of California Press.

Smith, Vardaman R. 1985. "John Stuart Mill's Famous Distinction between Production and Distribution." *Economics and Philosophy* 1:267–84.

Smyth, R. L., ed. 1962. *Essays in Economic Method.* London: Gerald Duckworth.

Sober, Elliott R. 1992. "Darwin's Nature." In *The Concept of Nature,* ed. John Torrance. Oxford: Clarendon.

Soper, Kate. 1995. *What Is Nature?* Oxford: Blackwell.

Sowell, Thomas. 1974. *Classical Economics Reconsidered.* Princeton, NJ: Princeton University Press.

Spary, Emma. 2000. *Utopia's Garden: French Natural History from Old Regime to Revolution.* Chicago: University of Chicago Press.

———. 2003. "'Peaches Which the Patriarchs Lacked': Natural History, Natural Resources, and the Natural Economy in France." In *Oeconomies in the Age of Newton,* ed. Margaret Schabas and Neil De Marchi. Durham, NC: Duke University Press.

Spengler, Joseph J. 1942. *French Predecessors of Malthus.* Durham, NC: Duke University Press.

———. 1984. "Boisguilbert's Economic Views vis-à-vis Those of Contemporary *Réformateurs.*" *History of Political Economy* 16:69–88.

Spiegel, Henry William. 1991. *The Growth of Economic Thought.* 3rd ed. Durham, NC: Duke University Press.

Stauffer, Robert Clinton. 1960. "Ecology in the Long Manuscript Version of Darwin's *Origin of Species* and Linnaeus' *Oeconomy of Nature.*" *Proceedings of the American Philosophical Society* 104:235–41.

Stephens, David W., and John R. Krebs. 1986. *Foraging Theory.* Princeton, NJ: Princeton University Press.

Steuart, Sir James. 1767/1966. *An Inquiry into the Principles of Political Oeconomy.* Edited by Andrew Skinner. 2 vols. Chicago: University of Chicago Press.

———. 1794/1966. "Biographical Sketch of Own Life." In *An Inquiry into the Principles of Political Oeconomy,* ed. Andrew Skinner. Chicago: University of Chicago Press.

Stewart, Balfour, and P. G. Tait. 1875. *The Unseen Universe; or, Physical Speculations on a Future State.* London: Macmillan.

Stewart, M. A., ed. 1990. *Studies in the Philosophy of the Scottish Enlightenment.* Oxford: Clarendon.

Stillingfleet, Benjamin, ed. and trans. 1791/1977. *Miscellaneous Tracts Relating to Natural History, Husbandry, and Physick.* 3rd ed. Reprint, New York: Arno. The 1st edition appeared in 1759.

Stroud, Barry. 1977. *Hume.* London: Routledge.

Sutton, Geoffrey. 1981. "Electric Medicine and Mesmerism." *Isis* 72:375–92.

———. 1995. *Science for a Polite Society: Gender, Culture, and the Demonstration of Enlightenment.* Boulder, CO: Westview.

Swinback, Peter. 1982. "Experimental Science in the University of Glasgow at the Time of Joseph Black." In *Joseph Black, 1728–1799: A Commemorative Symposium,* ed. A. D. C. Simpson. Edinburgh: Royal Scottish Museum.

Taylor, Charles. 1984. *Sources of the Self.* Cambridge: Cambridge University Press.

Teichgraeber, Richard F. 1986. *"Free Trade" and Moral Philosophy: Rethinking the Sources of Adam Smith's Wealth of Nations.* Durham, NC: Duke University Press.

Thomas, Brinley. 1991. "Alfred Marshall on Economic Biology." *Review of Political Economy* 3:1–14.

Thompson, William, and P. G. Tait. 1867. *Treatise on Natural Philosophy.* Oxford: Clarendon.

Torrance, John, ed. 1992. *The Concept of Nature.* Oxford: Oxford University Press.

Tribe, Keith. 1978. *Land, Labour and Economic Discourse.* London: Routledge.

———. 1981. *Genealogies of Capitalism.* London: Macmillan.

———. 1999. "Adam Smith: Critical Theorist." *Journal of Economic Literature* 37:609–32.

Turgot, A. R. J. 1770/1898. *Reflections on the Formation and the Distribution of Riches.* Translated by William J. Ashley. New York: Macmillan. The *Reflections* was written and circulated in 1766 but first published only in 1770.

———. 1970. *Écrits économiques.* Paris: Calmann-Lévy.

Vaggi, Gianni. 1987. *The Economics of Francois Quesnay.* London: Macmillan.

Vanderlint, Jacob. 1734/1914. *Money Answers All Things.* Reprint, New York: Johnson Reprint.

Vatin, François. 1998. *Économie politique et économie naturelle chez Antoine-Augustin Cournot.* Paris: Presses Universitaires de France.

Vaughn, Karen. 1980. *John Locke: Economist and Social Scientist.* Chicago: University of Chicago Press.

Veblen, Thorstein. 1899/1924. *Theory of the Leisure Class.* Reprint, London: George Allen & Unwin.

Vickers, Douglas. 1959. *Studies in the Theory of Money, 1690–1776.* Philadelphia: Chilton.

Vidal, Fernando. 1993. "Psychology in the Eighteenth Century: A View from Encyclopedias." *History of the Human Sciences* 6:89–119.

Viner, Jacob. 1972. *The Role of Providence in the Social Order.* Philadelphia: American Philosophical Society.

———. 1991. "Adam Smith and Laissez Faire." In *Essays on the Intellectual History of Economics,* ed. Douglas A. Irwin. Princeton, NJ: Princeton University Press.

Wallace, A. R. 1753. *A Dissertation on the Numbers of Mankind.* Edinburgh: Printed for G. Hamilton & J. Balfour.

Waterman, A. M. C. 1991. *Revolution, Economics and Religion: Christian Political Economy, 1798–1833.* Cambridge: Cambridge University Press.

———. 1996. "Why William Paley Was 'the First of the Cambridge Economists.'" *Cambridge Journal of Economics* 20:673–86.

Weintraub, Roy. 1991. *Stabilizing Dynamics: Constructing Economic Knowledge.* Cambridge: Cambridge University Press.

Wennerlind, Carl. 2001. "The Link between David Hume's *Treatise of Human Nature* and His Fiduciary Theory of Money." *History of Political Economy* 33:139–60.

Werhane, Patricia. 1991. *Adam Smith and His Legacy for Modern Capitalism.* Oxford: Oxford University Press.

Westfall, Richard S. 1958. *Science and Religion in Seventeenth-Century England.* New Haven, CT: Yale University Press.

Weulersse, Georges. 1910. *Le mouvement physiocratique en France (de 1756 à 1770).* 2 vols. Paris: Mouton.

Whately, Richard. 1832. *Introductory Lectures on Political Economy.* Dublin.

Whitaker, John. 1975. *The Early Economic Writings of Alfred Marshall, 1867–1890.* 2 vols. London: Macmillan.

———. 1977. "Some Neglected Aspects of Alfred Marshall's Economic and Social Thought." *History of Political Economy* 9:191–97.

———, ed. 1990. *Centenary Essays on Alfred Marshall.* Cambridge: Cambridge University Press.

———, ed. 1996. *The Correspondence of Alfred Marshall, Economist.* Vol. 1, *Climbing, 1868–1890.* Cambridge: Cambridge University Press.

White, Michael V. 1991. "Frightening the 'Landed Fogies': Parliamentary Politics and *The Coal Question.*" *Utilitas* 3:289–302.

———. 1994a. "Bridging the Natural and the Social: Science and Character in Jevons's Political Economy." *Economic Inquiry* 32:429–44.

———. 1994b. "The Moment of Richard Jennings: The Production of Jevons's Marginalist Economic Agent." In *Natural Images in Economic Thought,* ed. Philip Mirowski. Cambridge: Cambridge University Press.

Wicksteed, Philip Henry. 1910/1933. *The Common Sense of Political Economy and Selected Papers and Reviews on Economic Theory.* 2 vols. Rev. ed. Edited by Lionel Robbins. London: George Routledge & Sons.

Wightman, W. P. D. 1980. Introduction to *Essays on Philosophical Subjects,* by Adam Smith, ed. W. P. D. Wightman. Oxford: Clarendon.

Wilde, C. B. 1982. "Matter and Spirit as Natural Symbols in Eighteenth-Century British Natural Philosophy." *British Journal for the History of Science* 15:99–131.

Williams, Raymond. 1976. *Keywords: A Vocabulary of Culture and Society.* New York: Oxford University Press.

Wilson, Benjamin. 1746. *An Essay towards an Explication of the Phenomena of*

Electricity, Deduced from the Aether of Sir Isaac Newton. London: Printed for
C. Davis & M. Cooper.

Wilson, David B. 1977. "Concepts of Physical Nature: John Herschel to Karl Pearson." In *Nature and the Victorian Imagination,* ed. U. C. Knoepflmacher and
G. B. Tennyson. Berkeley and Los Angeles: University of California Press.

Wilson, Fred. 1990. *Psychological Analysis and the Philosophy of John Stuart Mill.*
Toronto: University of Toronto Press.

Winch, Donald. 1978. *Adam Smith's Politics: An Essay in Historiographic Revision.*
Cambridge: Cambridge University Press.

———. 1996. *Riches and Poverty: An Intellectual History of Political Economy in
Britain, 1750–1834.* Cambridge: Cambridge University Press.

Wise, M. Norton. 1993. "Mediations: Enlightenment Balancing Acts, or the Technologies of Rationalism." In *World Changes: Thomas Kuhn and the Nature of
Science,* ed. Paul Horwich. Cambridge, MA: MIT Press.

———, ed. 1995. *The Values of Precision.* Princeton, NJ: Princeton University Press.

Wise, M. Norton, with Crosbie Smith. 1989–90. "Work and Waste: Political Economy and Natural Philosophy in Nineteenth Century Britain." *History of Science*
27:263–301; 27:391–449; 28:221–61.

Wood, Gordon S. 1982. "Conspiracy and the Paranoid Style: Causality and Deceit in
the Eighteenth Century." *William and Mary Quarterly* 39:401–41.

Wood, Paul B. 1989. "The Natural History of Man in the Scottish Enlightenment."
History of Science 27:89–123.

———. 1990. "Science and the Pursuit of Virtue in the Aberdeen Enlightenment." In
Studies in the Philosophy of the Scottish Enlightenment, ed. M. A. Stewart. Oxford: Clarendon.

———, ed. 2002. *Essays and Observations, Physical and Literary, Read before a Society in Edinburgh.* 3 vols. Bristol: Thoemmes Continuum. The three volumes
initially appeared over the period 1754–71.

Worster, Donald. 1977. *Nature's Economy: A History of Ecological Ideas.* Cambridge:
Cambridge University Press.

Wynne, Brian. 1979. "Physics and Psychics: Science, Symbolic Action, and Social
Control in Late Victorian England." In *Natural Order: Historical Studies of Scientific Culture,* ed. Barry Barnes and Steven Shapin. Beverly Hills, CA: Sage.

Xenophon. *Oeconomicus.* Translated by Sarah B. Pommeroy. Oxford: Clarendon.

Young, Jeffrey T. 1990. "David Hume and Adam Smith on Value Premises in Economics." *History of Political Economy* 22:643–57.

Young, Robert M. 1985a. "Darwinism *Is* Social." In *The Darwinian Heritage,* ed.
David Kohn. Princeton, NJ: Princeton University Press.

———. 1985b. *Darwin's Metaphor: Nature's Place in Victorian Culture.* Cambridge:
Cambridge University Press.

Zemagni, Stefano. 1989. "Economic Laws." In *The Invisible Hand: The New Palgrave,* ed. J. Eatwell, M. Milgate, and P. Newman. New York: Norton.

INDEX